BEST BORDERS

BEST BORDERS

TONY LORD

LONDON NEW YORK SYDNEY TORONTO

This edition published 1994 by BCA
by arrangement with Frances Lincoln Limited
4 Torriano Mews, Torriano Avenue
London NW5 2RZ

CN 3403

Best Borders

Set in Adobe Garamond 10½ on 13pt

Printed and bound in Italy by New Interlitho S.p.A.

First Frances Lincoln edition April 1994

*To my mother, Marjorie Lord, who introduced
me to gardening.*

ABOVE *The Top Border at The Priory, Kemerton, Hereford
and Worcester (see page 86) was first planted by Peter and
Elizabeth Healing in 1957. Every colour of the rainbow was
included, even hot colours, unfashionable for many years but
now recognized as invaluable, as are the dahlias (here* D.
'Blaisdon Red'), on which much of the display depends.

HALF-TITLE PAGE *At Great Dixter in East Sussex (see page
96), Christopher Lloyd has created an inspired and
idiosyncratic border: a series of incidents shows daring use
of colour and bold plant associations. Here magenta*
Geranium *'Ann Folkard', supported on brushwood,
contrasts with* Euphorbia schillingii *and* Achillea *'Taygetea'.*

FRONTISPIECE *The Purple Border at Sissinghurst Castle in
Kent (see page 62) has been refined and enhanced by a
remarkable team of gardeners since it was first planted by
Vita Sackville-West. It demonstrates how best to use a
difficult colour range and how to extend the display
upwards by means of a wall.*

CONTENTS

Introduction 6

FORMAL DOUBLE BORDERS 14
The twin herbaceous borders at Arley Hall, Cheshire: generous twentieth-century planting in a magnificent Victorian setting.

A BORDER IN THE 'MINGLED' STYLE 28
The Wall Border at Packwood House, Warwickshire: closely packed planting for abundant effect for many months.

DOUBLE SUMMER BORDERS 36
Fin de siècle magnificence in the Wall Garden at Nymans, West Sussex: annuals and tender plants backed by tall perennials.

RED DOUBLE BORDERS 44
The borders created by Lawrence Johnston at Hidcote Manor in Gloucestershire: mixed planting with bulbs and tender perennials in a clashing colour range from orange to crimson against dusky foliage.

BORDERS WITH ANNUALS 54
Borders at Le Pontrancart, Normandy, first designed in the 1930s by Kitty Lloyd-Jones.

A PURPLE BORDER 62
The border at Sissinghurst Castle, Kent, created by Vita Sackville-West: a mixed border backed with wall plants, using temporary planting.

CONTRASTING DOUBLE BORDERS 76
Formal borders by Graham Thomas at Cliveden, Buckinghamshire: strong vibrant colours face the opposite border of misty pale flowers.

A RAINBOW BORDER 86
The Top Border created by Peter Healing at The Priory, Kemerton in Hereford and Worcester: a spectrum of colours with stunning use of hot tones and dark foliage.

BOLD COLOUR IN A MIXED BORDER 96
Christopher Lloyd's Long Border at Great Dixter, East Sussex: surprising juxtapositions of colour that are attractive all year round.

INFORMAL EARLY SUMMER BORDERS 106
Hidden borders by George Smith at the Manor House, Heslington, York: exuberant and informal planting at its peak in early summer.

BORDERS IN A SMALL TOWN GARDEN 116
Anne Dexter's garden in Oxford: skilful planting demonstrating how to make the most of a narrow site.

BORDERS WITH TENDER PERENNIALS 126
The West Court Borders by Tony Lord at Hardwick Hall, Derbyshire: rich colours arranged in Jekyllian progression for maximum impact from midsummer to autumn.

Gardens to visit 138   Further reading 138   Index 139
Hardiness zones 143   Acknowledgments 144

INTRODUCTION

Many gardeners today consider there is a single classic and universal style of border, derived from the writings of Gertrude Jekyll and exemplified by the planting of gardens such as Hidcote or Sissinghurst. This is simplistic in the extreme: it denies both the diversity of Miss Jekyll's designs and the individuality of today's gardens, which have evolved far beyond her teachings to demonstrate very varied styles of their own. *Best Borders* aims to show just how different twelve borders can be, to analyse why they succeed and discuss what practical techniques contribute to their excellence.

Herbaceous borders have existed for only about 150 years, those at Arley Hall in Cheshire being perhaps the earliest in Great Britain. Gardeners now see border planting largely through the eyes of Gertrude Jekyll. But Miss Jekyll's is not the only style; a number of daring and innovative garden designers have since produced variations on her theme which have made quite different use of form and colour and which, although seldom recognized as such, are utterly different in concept.

Though borders have long been an important element of gardens, the border of today in which plants are massed for a large-scale effect is a relatively recent phenomenon. The word itself was first applied to beds around the fringe of a garden in medieval times; perhaps the fact that early examples were raised and edged with boards gave added meaning to the word. Into the eighteenth century, borders were places where flowers were grown for close inspection, set apart from each other in what is called 'sparse planting'. Such borders were utterly different from those of today in that they were not created for large-scale effect, to be appreciated as a complete eyeful of intergrading colours and textures of both flowers and foliage, but were simply places where flowers were grown so that the blooms of each could be examined at close quarters.

Plantsmen of the seventeenth century not only grew plants in borders around the edge of the garden but made elaborate geometric patterns of beds called frets. Gardeners such as John Rea (d. 1681) gave precise prescriptions of how the borders and frets were to be made and how the plants were to be arranged within them: borders around the walls were to be 1.5m/5ft broad and the beds within the fret 75cm/30in. The wall at the back of the border would be planted with the choicest fruit, an important element of seventeenth-century gardens. Rea recommended planting the border with auriculas, red primroses, hepaticas, double rose campion, double meadow saxifrage, double rocket, the best wallflowers and double stocks, with crocuses along the front edge. He preferred to grow different plants in the beds of the fret, such as crown imperials, lilies, 'great Tufts of the best *Pionies*', daffodils, hyacinths, tulips, ranunculus, anemones, fritillaries and bulbous iris. The flowers were usually varieties of Old World

OPPOSITE Plates bandes *at Het Loo near Appeldoorn in the Netherlands demonstrate how completely styles of border planting have changed since the late seventeenth century. Save for topiary yew and box, the plants play a subsidiary role in garden design, their flowers providing the merest ornamentation to an architectural and geometric design. Some consider this excess of formality an attempt to show man's ability to dominate nature, something so self-evident today that there is no longer a need to express it. Here crown imperials and parrot tulips are used much as John Rea recommended.*

LEFT *Phyllis Reiss created some excellent and influential borders in her own garden at Tintinhull and here at Montacute, both in Somerset. Strong colours were used to complement warm-coloured stone and stand up to the powerful architecture of the house and the garden walls. Stark white flowers, here* Clematis recta, *were included to leaven purple, red and gold, which used alone would seem too heavy. A limited range of plants was repeated at regular intervals to give rhythm and unity. Smoke bushes add bulk with colour from* Penstemon *'Andenken an Friedrich Hahn' (P. 'Garnet'),* Thalictrum flavum *subsp.* glaucum, *salvia and achillea.*

species, selected for their extraordinary stripes, picotee petals or double flowers. Rea warned that box edging can remove goodness from the soil, though he recommended it for large beds in the fruit garden.

During the seventeenth century, parterres of elaborate box scrollwork became popular in the grandest gardens, often edged with narrow borders or *plates bandes* similar in planting to Rea's frets. Examples of these can still be seen at William and Mary's palace at Het Loo near Appeldoorn in the Netherlands. By the end of the century, box was used in great quantity for edging as well as scrollwork and choice and immensely expensive varieties of flowers such as tulips, carnations and auriculas would form the basis of the display; save for a few annuals, there were not many flowers for late summer. Most of the plant introductions from overseas had yet to reach gardens and the range that was available was

simply not capable of imitating today's borders.

The arrival of the informal landscape style during the eighteenth century produced gardens which had no place for formal borders; flowers were usually banished to a walled flower garden. Doubtless these were liberally supplied with borders, though the fact that representations of them are rare suggests that their design was not a priority in garden aesthetics. The creation of a flower garden by the poet William Mason for the Earl of Harcourt at Nuneham Park, Oxfordshire, begun in 1771, was an important break: not only did it once more make flowers a key element of garden design, but it used them in a large-scale way. Beds in amoeboid shapes and planted thickly with flowers lined the vistas of the garden. For the first time, a solid mass of varied herbaceous planting, whole plants and not just flowers, was used to make a grand effect.

Beds such as these, which we would call island beds today, were popular in the early nineteenth century, even some time after the landscape gardener Humphry Repton (1752–1818) had repopularized formal flower gardening near the house, bringing back the border as an element of grand garden design. In both borders and informal beds, colours were thoroughly mixed so that no two varieties of the same colour were placed next to each other. But in a century which began with Repton and ended with Jekyll, it was perhaps the ever-increasing availability of species from abroad and the variety of new hybrids that was the most important factor in allowing the further refinement and development of the border.

The introduction of new plants had been hampered by several factors: on long sea journeys, precious drinking water could not be spared for plants; those that survived the

journey tended to be succulents, dormant bulbs or tubers or species with long-lived seed. Faster sea crossings and the invention of the Wardian case, a portable greenhouse which conserved moisture and eliminated the need for watering, made it possible to introduce a host of new species. Plant hunters were hired to scour the continents for good garden plants. David Douglas (1799–1834) introduced to gardens many annuals, perennials, shrubs and trees from the West Coast of North America. Political factors had also hampered new introductions, several important areas being closed to foreigners. The opening of China in 1842 and Japan in 1858 allowed yet more species to be introduced.

With a growing understanding of genetics, the hybridists were able to use this host of new species to produce countless thousands of

improved varieties. Many of these newly arrived plants, particularly those derived from prairie and meadow flowers capable of competing with other plants of their own height, proved suitable for borders. The assembling of so many thousands of varieties of garden plants made it possible to specialize. For the first time, gardeners could have early summer borders, late autumn borders, annual borders or blue borders.

Gardeners came to recognize the value of tender perennials for continuous bloom and annuals for late display, only to suffer the wrath of the irascible William Robinson (1838–1935) in books such as *Hardy Flowers* (1871) and *The English Flower Garden* (1883). Robinson considered such plants bloomed for an indecently long time and needed too much work; he insisted they

should on no account be mixed with worthy herbaceous perennials and believed that 'no plan which involves expensive yearly effort on the same piece of ground can ever be satisfactory'. It is fortunate that many gardeners chose to ignore this dogmatism. Robinson was perhaps more influential as a publisher of horticultural books and journals than as a practitioner of gardening; his ideas lacked both the design flair and the thoughtful balance of Gertrude Jekyll's writing.

There can be no doubt that the books of Gertrude Jekyll (1843–1932) contain a wealth of aesthetic and practical common sense on the planting of gardens, as valuable today as when they were written. She was undoubtedly the first garden writer to show how to exploit fully the use of colour, foliage texture and form to achieve a co-ordinated

LEFT *The D-shaped herbaceous garden at the National Trust's Anglesey Abbey in Cambridgeshire is grand in scale and excellent in the cultivation of lawns, hedges and border plants. The planting is simple and unforced with no colour scheme, except for the avoidance of the worst clashes. Its immediate impact and charm are undoubted. We should not necessarily consider the lack of Jekyllian use of colour as a fault, though private gardeners, with limited lengths of borders, will often prefer to wring every last scrap of aesthetic effect from their own gardens by controlling colour more.*

RIGHT *Alan Bloom's island beds at Bressingham, Norfolk, designed to give the best conditions for an outstanding collection of plants, have influenced many gardeners, myself included. However, to all but the most ardent plantsmen, allowing the plants to shape the garden to such an extent might seem like letting the tail wag the dog; gardeners will often prefer to tailor the plants to the design rather than vice versa.*

and sophisticated result, as effective at close range as from a distance. However, her style has limitations: she used a restricted range of plants which, though precisely and skilfully employed, would not satisfy those gardeners of today who want to grow a wide range of plants; her designs often provided a display lasting only a few weeks, not a problem in large gardens of many 'rooms' each with a different season but limiting for those of us who have only one border which must last in beauty for most of the year. Her planting often demanded far more labour than most home gardeners could provide.

Miss Jekyll's writings, even in her own day, represented only the visible tip of a substantial iceberg of horticultural opinion. Arthur Hellyer (1902–92), for long the doyen of British gardening writers,

considered that she was not a great innovator but merely wrote about what many gardeners were already doing. She did *not* write about other attractive styles of gardening which were equally prevalent in her day and which she probably took for granted. Because her views have become so accepted, it is easy to forget that other styles existed but in this book there are two survivals, Nymans and Packwood, both of them with considerable charm though utterly un-Jekyllian.

It is absurd to explain twentieth-century planting solely in terms of Miss Jekyll's writings, although this practice seems to be prevalent. We would not consider all early nineteenth-century music to be written in the manner of Beethoven, nor would we expect compositions written sixty years after his death to be simply a reworking of his style.

Gardening, like music, has always been diverse and will always evolve. No one figure, however great, will predominate utterly, not even in his or her own day and certainly not a century later.

True, some gardeners consider themselves to be guided principally by Miss Jekyll's writings. But is this really so? When we devise a planting scheme, most of us are influenced by all the gardens we have seen and by all the garden writers we have read, not to mention Aunt Maud and Mrs Smith down the road. Even the most influential of them – Gertrude Jekyll, William Robinson, Graham Stuart Thomas or Aunt Maud – accounts for at most only ten per cent of our inspiration. Though our influences are so diverse, this need not be a recipe for a characterless mish-mash; there is still room for inspired

9

The double borders in the centre of the Old Rose Garden at Mottisfont Abbey in Hampshire, planned by Graham Thomas for the National Trust, provide interest from spring until autumn. Starting with bold foliage of cardoons, irises and bergenias, summer-flowering plants are followed in autumn by asters, kniphofias and standard Hibiscus syriacus. *Plant groups are generous in scale and mostly low in height. Here* Alchemilla mollis *and* Stachys byzantina *spill on to the path in front of cream* Sisyrinchium striatum. *Such planting creates a spacious feel and allows uninterrupted views across the borders into the four quarters of the garden, unusual in Britain where a sense of enclosure is almost a prerequisite and most borders are backed by a hedge or wall.*

Charles Cresson's garden at Hedgleigh Spring, Pennsylvania, is a rare example of an early twentieth-century American flower garden, and illustrates American rather than European characteristics. For instance, the picket fence gives only partial enclosure and allows a view across the border; and a greater proportion of plants derives from North American species. Maintaining such differences helps make borders seem to belong.

gardeners to make a new synthesis quite different from anything that has been made before, as some of the examples in *Best Borders* show.

The story does not end with Miss Jekyll. World War II caused some significant changes in the way borders were planted. Professional gardeners became scarcer and more expensive; labour-saving styles became predominant. Tender perennials almost disappeared and annuals went into a decline from which they are only now beginning to emerge. In the years since the war, Graham Thomas, Gardens Adviser to the National Trust from 1956 until his retirement in 1974 and an admirer of Miss Jekyll (though his own style has significant differences), has been an important influence, having designed borders of great quality in numerous Trust gardens.

Plantsman and nurseryman Alan Bloom has championed island beds, created in his garden at Bressingham Hall in Norfolk in the years since World War II as an alternative to the traditional border. These differ from island beds of the late eighteenth and early nineteenth centuries only in their larger groups and wider range of plants. Their advantages are that they allow access of air and light all around each plant group, encouraging healthier growth and minimizing the need for staking. This undoubtedly benefits the plants, cuts down the work of staking and makes for more natural growth, though the very informality of the beds makes it difficult to define areas of the garden and give focus. Though they are suited to the larger garden, in smaller sites it is hard to make island beds harmonize with clear rectangular boundaries.

Through the middle years of the century, the German nurseryman Karl Foerster (1874–1970) encouraged rich herbaceous planting, popularizing also ferns and grasses, used in altogether a more naturalistic way in open swathes and beds, again moving away from the concept of the traditional herbaceous border. Much of the current popularity of grasses on both sides of the Atlantic can be considered to be his legacy, the brilliant use of them in designs for North American gardens by Oehme and van Sweden being a particularly eximious example.

Even since the 1960s there have been some clear trends in border planting. Championed in the British Isles by gardeners such as Mrs Desmond Underwood, silver foliage became immensely popular, used not only with pinks, blues and mauves but also as a foil to brighter colours such as magenta. By the 1980s, many gardeners had had enough of the suffocating softness of such schemes and more vibrant colours such as scarlet, orange and yellow returned to favour. It is possible now to detect a shift in favour of more daring colour schemes using two or more contrasting colours, purple and salmon, scarlet and chartreuse, blue and yellow, these being probably more popular than single-colour or purely Jekyllian schemes.

The borders in this book show styles of planting dating from the 1840s to the present day, each example being taken in the order of the date of its style. Thus the Yellow Border at Packwood, a twentieth-century recreation of the early nineteenth-century 'mingled' style, features early in the book. This chronological treatment is intended to help show how borders have evolved and to highlight how far the later examples have progressed beyond Miss Jekyll, even beyond their original planting.

The twelve examples are chosen to give a wide range of flowering seasons, colours and plant content, some being purely herbaceous while others include shrubs or annuals. Underplanting with spring bulbs to further lengthen the season of display, almost an essential embellishment, particularly for gardeners with just one border, is demonstrated at Great Dixter, though several other featured borders have been planted with tulips or other spring flowers.

To seem to fit, it helps if gardens respect historic and local traditions of planting as well as the personal likes of the owner. In several of the examples in this book, a respect for past styles of planting has served to give character and a sense of belonging. None of the borders has been devised solely by sketching a plan on to paper and considering the job finished. Even the most skilled designers will admit that planting needs to be honed to perfection over several growing seasons, often a decade or more, adjusting

colours and textures until they are just right and replacing plants which do not live up to expectations.

Some borders here, such as those at Arley Hall, have evolved over several generations of a single family. Some, first made by famous gardeners such as Vita Sackville-West (Sissinghurst) or Lawrence Johnston (Hidcote), have continued to evolve under a succession of gardeners, becoming richer and better than their originators might ever have thought possible, while remaining absolutely true to their original concept. Others have been gardened solely by their creators. In every case, there has been an understanding of what was required, a steadfastness of purpose, constant

criticism, adjustment and improvement.

It has always been difficult to find hardy plants suitable for a late summer border. From the late nineteenth century until World War II, annuals often filled this role, though their popularity declined rapidly during the war. Though we now recognize that annuals are useful, in spite of a modicum of extra work, many of the best varieties have disappeared, particularly single-colour strains in heights other than the very dwarf. One hopes that the best of these can be brought back from the brink of extinction, especially the tall antirrhinums so beloved of Miss Jekyll but now so hard to find. Tender perennials such as salvias, cannas, dahlias and

argyranthemums, also much used before the war for late colour, have enjoyed a heartening return to popularity in recent years. Three of the borders here, Nymans, Le Pontrancart and Hardwick, are designed for the latter part of the year and make extensive use of annuals or tender perennials.

That so many of the borders in this book should be in England is partly coincidence and partly a matter of the author's convenience. Survivors of nineteenth-century styles of planting might be scarce outside the British Isles but otherwise a similar range of borders could be found in the gardens of France, Germany, North America or many other areas. Nor is there anything exclusively British about the principles of planting, colour, foliage texture and form shown in the borders, all of which apply universally and could be adapted to any hardiness zone from Z4 to Z11. The grand scale of the gardens offers scope to the author and photographer to illustrate important points with sufficient examples but should not

discourage those with more modest plots. Anne Dexter's garden shows just how much pleasure borders in a small garden can give and almost every example in *Best Borders* could be scaled down to fit the tiniest garden, or even a window box.

Inevitably there is not enough room in such a book to include many favourite borders. I would have liked an example which showed grasses, even possibly used alone, a harmony of similar textures in shades of green. I admire immensely the borders created at Broughton Castle in Oxfordshire: one a confection of pinks, whites and blues in early summer; another, smaller, border of annuals and cottage garden flowers jumbled together intimately in the most charming confusion. Then there is Frank Cabot's flower garden at Stonecrop, New York, a lattice of beds divided by narrow grass paths filling the entire garden, the colours shifting gradually from bed to bed, uninterrupted by the paths; this seems to overturn the linear nature of the border, widening it to fill

another dimension. But it is perhaps invidious to single out more examples; there are no doubt enough other borders of equal quality and individuality to fill several more volumes.

Photographing the gardens myself has proved a useful discipline. The photographer sees both triumphs and flaws which might easily be overlooked by the gardener or writer; analysing why a photograph works can reveal some of the secrets of good planting. The flaws can be all too obvious through the lens, though we might not think to correct such defects in our own garden: obtrusive labels or stakes, ugly deadheads, swathes of flowerless plants disrupting the flow of colours, borders placed where sunlight flatters them least. The triumphs are often harder to analyze. Why does a picture of plants grouped together have so much more beauty than the individual plants alone? Is it the interplay of colours or shapes, the bold contrast of foliage textures or the way the plants are lit by the sun? The answers to such questions are the key to inspired gardening.

RIGHT *Against one wall of the Ladies' Garden at Broughton Castle, Lord and Lady Saye and Sele's home in Oxfordshire, a charming mix of cottage-garden flowers looks for all the world as though they had placed themselves with scarcely any interference from the gardener. This is an illusion, for such apparent wildness is the most transient of states in gardens unless balance is maintained with sensitivity, skill and a good deal of work. Though the planting of annuals, biennials and perennials changes from year to year, the effect of delightful informality remains the same.*

OPPOSITE *The main border along the Long Walk at Broughton Castle is at its best when the roses (here 'Fantin-Latour') are in full bloom. Differing from the borders at the Manor House, Heslington (see page 106), also planted for early summer, in its greater formality and paler colours – including a good deal of white – the effect is no less successful though quite dissimilar. Roses are complemented with delphiniums, campanulas and* Crambe cordifolia, *the stone edging softened by pinks, geraniums and* Allium cernuum.

FORMAL DOUBLE BORDERS

Arley Hall

The twin borders at Arley Hall, set in the lush and gentle greenness of the Cheshire countryside, seem the archetype of the traditional herbaceous border, magnificent in scale and timeless in planting. Some say that these borders are the oldest in existence; others argue that those at Newstead Abbey, Byron's family home in Nottinghamshire, are older. Whatever the truth, it is hard to imagine how such a scheme could be more perfectly scaled or more splendidly proportioned.

Both borders at Arley are of generous width, so much so that one scarcely notices that they are not equal. Such ample depth allows about six ranks of plant groups of good size; this gives an extra richness and the possibility for more interesting interactions between groups than a border of more usual size only three or four groups deep, akin to the advantages of an orchestra a hundred strong over one of a mere fifty players.

It is not just the size of the Arley borders that makes them so impressive: the framework of the wall on one side and the yew hedges on the other, both with periodic punctuation of

buttresses of yew, gives a satisfying structure lacking from some other twin borders of similar scale. Yew is supreme for providing a formal setting: its fine, even texture gives the potential for crisply finished architectural form and its rich colour is the perfect foil for flowers. The degree of ornamentation, often carried to excess by the Victorians, seems perfectly judged here and the summerhouse at one end and parkland at the other provide agreeable focal points from whichever direction the borders are viewed.

The size and structure of the Arley borders have always been impressive. The planting orchestrated by Lady Ashbrook since the 1960s has transformed them into a magnificent display with a distinct colour scheme: in the early part of the year, until midsummer, the borders are a cool, magical and misty mix of blues, soft pinks, greys and whites seasoned with acid yellow; from late summer until the frosts, the hot colours, yellow, orange and red, take over.

The garden began to take shape when in 1786 Sir Peter Warburton and his young wife

The sunnier of the twin borders at Arley contains plants that enjoy a summer baking and would grow less well on the shadier side; these include Allium giganteum *and* A. christophii, *alstroemerias and fiery* Crocosmia *'Lucifer'. Softer colours seem further away, exaggerating the perspective. Ramrod stems of*

*orach (*Atriplex hortensis var. *'Rubra'), still slender before the burden of autumn seed, and blue-green poppy pods provide repeated accents.*

employed William Emes to landscape the grounds. The suggestion that Emes was a pupil of 'Capability' Brown is without proof, though after Brown's death he followed his smoothly naturalistic style of landscaping. However, unlike Brown, he seems to have been one of the first designers to repopularize flower gardens immediately adjacent to the house: at Sandon in Staffordshire in 1781 he had proposed a flower garden beneath the windows of the new dining-room and he advised a backing of flowering shrubs for a flower border.

The 1786 plan of Arley, showing either what already existed or Emes's first proposals, features three flower gardens with island beds in the walled enclosures around the house; shrubs and trees are visible in one of these but there is no trace of formal borders. The borders might have followed as early as 1791, though this would be unusual for the period: Humphry Repton's work, which revived the use of formal elements of the garden, would not by then have been widely copied, and the term 'herbaceous border' was not to appear in print until John Claudius Loudon's *Encyclopaedia of Gardening* in 1822.

The term was used by writers such as Loudon and George Johnson in the 1820s and '30s to describe the sparse planting favoured by seventeenth-century gardeners like John Rea. Repetition at intervals of choice specimens surrounded by bare earth was seen as an old-fashioned ideal, allowing appreciation of the beauty of individual plants. Loudon maintained that 'Flowers in borders should always be planted in rows, or in some regular form, and this appearance should be assiduously kept up by trimming off all irregular side-shoots and straggling stalks, and reducing the bulk of plants which grow too fast. Every approach to irregularity, and a wild, confused, crowded, or natural-like appearance, must be avoided in gardens avowedly artificial.' So there.

Sir Peter had died in 1813 and been succeeded by nine-year-old Rowland Egerton-Warburton. Even after his coming of age, young Rowland was constrained by lack of funds from developing the garden further; the rebuilding of the house in 1832 had to be financed by cutting down woods and the building of the chapel had to wait for an inheritance in 1839.

The borders first appear in a plan of 1846 but it was not until 1856 that they took their present form. Then, Rowland's brother-in-law James Bateman, creator of one of the most remarkable Victorian gardens at Biddulph Grange, gave advice on planting yew hedges and buttresses for the borders, saying that they must be planted in very early autumn with 1.2m/4ft high plants 60cm/2ft apart, taking out every other plant after about three years. Bateman's advice is still good today, though we would generally be too timid and short-sighted to remove half the plants; nevertheless, the spacing of 1.2m/4ft does ensure that no plants are crowded out, leaving unsightly brown patches or gaps as odd ones die. The one drawback to such wide spacing is that when hedges are restored by cutting back, the space will take much longer to fill.

The disadvantage of hedges as architecture is that unlike stone or brick, their shape changes, usually for the worse, as the plants grow. Gardeners always try to remove the annual extension of growth when clipping each year but inevitably some remains, making the hedge fatter, even if only by a thumbnail width or so per annum. At Arley the yews are now as much as 1.8m/6ft through, interfering with proportions and creating considerable extra work, though their additional bulk has a splendid solidity.

The great advantage of hedges over brick or some sorts of stone is that their colour is generally a more flattering background for flowers – and, of course, they are much cheaper to install.

Ideally, it helps to leave a gap between the back of the plants and the hedge, to make access easy for clipping or weeding and to ensure that both plants and hedge have plenty of light and air for optimum growth. The principle holds good even on a smaller scale.

A good hedge represents an investment for the future: thorough preparation is prudent for assets to be realized. I remember digging a trench 1.8m/6ft deep by 1.2m/4ft wide for a particularly important yew hedge in Roundhay Park, Leeds, an exercise that my fellow gardeners considered *de trop*, though they expressed their opinions more vigorously. So it proved to be: the compost with which the trench was filled subsided, taking the yews below the water table; they quickly turned brown, for yews are the most intolerant of plants, resenting being planted at any depth other than that at which they originally grew in the nursery.

But given reasonable preparation, good drainage, adequate light and regular feed with a balanced fertilizer, yew should grow at a rate of about 30cm/12in each year (much the same for holly, a little more for beech and hornbeam, a little less for box), a rate that should be enough even for impatient gardeners. True, faster-growing subjects such

RIGHT, ABOVE *Magnificent scale, good proportions and the uncomplicated but telling ornamentation of shaped yew buttresses and cakestand finials show the effectiveness of simple design. Save for the occasional flaming scarlet poppy, blue flowers and silver foliage predominate in early summer with touches of mauve and pale yellow.* Berberis thunbergii *'Atropurpurea Nana', the only shrub permitted, lends solidity and contrast of foliage colour to both borders.*

RIGHT, BELOW *The sunnier border runs along the outer wall of the walled garden while its twin is backed by a yew hedge that separates it from the Ilex Avenue and Rose Garden. Matching yew buttresses on both sides make equal compartments. The plan on page 26 is of the second bay seen on the left in the photograph above. A summerhouse gives a view down the borders' length towards lush parkland. Though it would be churlish to fault them for being so good, the borders have the one disadvantage that no other part of the garden can quite match their splendour.*

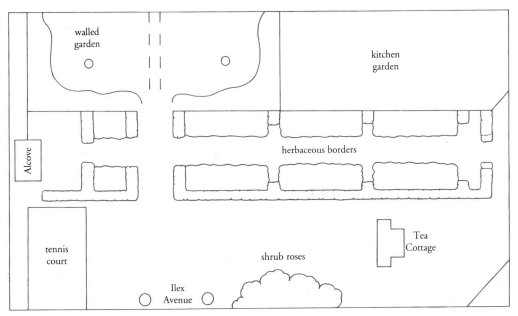

OVERLEAF, LEFT *The shadier border after an early summer shower. Purple-blue delphiniums bejewelled with raindrops and* Geranium × magnificum *impose the principal flower colour. Pale primrose blooms of giant scabious (*Cephalaria gigantea) *add gentle contrast, their simple shapes benefiting from a background of plain green yew. Topiary buttresses and finials add just enough punctuation and architecture, mimicked by a distant fir.*

OVERLEAF, RIGHT *Hot colours at the onset of autumn.* Montbretia (Crocosmia × crocosmiiflora), *vibrant against magenta-purple phlox, allows bold foliage contrast echoed by more crocosmias. Maroon leaves of orach, now weighted down with seed, and* Berberis thunbergii *'Atropurpurea Nana' unify a combination which would scarcely work with green leaves alone. Aconitum, rudbeckia, lythrum and dahlias add to the floral display.*

as Leyland cypress can make attractive and fine-textured hedges but they are harder to keep in good order in the long term and cannot easily be restored once they have grown out of shape.

It is an oft-repeated rule that good hedges should have a batter, that is they should slope inwards towards the top, though most gardeners probably prefer to see hedges with perfectly straight sides. It is certainly true that the narrower the top of the hedge, the denser the sides are likely to be; the broader the top, the greater the influence of apical dominance, whereby the uppermost shoots secrete hormones that suppress the sprouting of shoots lower down.

Provided yews are growing vigorously, it is possible to cut them back to the main stems to regrow to the original height and width; this method is reliable as far north as Cheshire but more risky in cooler climes. Tommy Acton, who gardened at Arley from 1939, retiring as Head Gardener in 1993, wishes that he had tackled the Arley hedges in the 1960s, before the garden was opened to the public; a couple of years of unsightliness then would have rejuvenated the hedges for another hundred years, as well as reducing by about thirty per cent the annual task of clipping them.

Other hedging shrubs used as a backing for borders can be treated in the same way, though holly, box, beech and hornbeam will sprout better from side branches than from the main stem. All of these retain leaves, whether dead or alive, in winter, appearing solidly architectural in every season. Of evergreen hedging shrubs for temperate climates, yew is perhaps the most versatile, though fastigiate (upright-growing) sorts such as Irish yew do not intermesh and should be avoided. European yew (*Taxus baccata*) is hardy only to Zone 6; for colder climates, *Taxus cuspidata* or *T. × media* cultivars will withstand Zone 5 conditions. Gardeners in still colder areas wanting evergreen hedges

can use Canadian hemlock, *Tsuga canadensis.*

European holly (*Ilex aquifolium*) or Highclere holly (*I. × altaclerensis*) (both Zone 7) would make a sparkling and handsome hedge, a magnificent background for most flowers, although gardeners would eternally curse its prickles: hand-weeding would be an unpleasant task. Spineless cultivars such as *Ilex aquifolium* 'Scotica' do not have such problems but are unaccountably seldom used and rarely grown in quantity by nurseries. The American holly (*Ilex opaca*) is hardier (Zone 6), though its leaves lack the glitter of the European sorts.

European box (*Buxus sempervirens*, Zone 6) grows rather more slowly. It is as fine as yew though a lighter colour and not capable of making a firm hedge of such height. Upright sorts such as 'Hardwickensis' and 'Handsworthiensis' seem to produce much stiffer stems and so are capable of making a fairly rigid hedge even though they do not intermesh well; however, I find their dull, dark grey-green foliage rather depressing and prefer richer green sorts such as 'Notata', often found under its descriptive synonym 'Gold Tip'. Gardeners wanting a particular shade of green as a foil for a colour-schemed border might prefer a clone with a bluish or yellowish cast, while some might opt for the range of subtly different shades of green provided from seed-raised plants, though these are hard to find and slow to raise oneself. Seed-raised yew also provides an attractive range of tones of green, though with less variation than seed-raised box, the difference between the individual plants being especially noticeable as the new shoots appear, tinged with bronze, gold or rusty red.

Buxus microphylla is a little hardier (Zone 5) than European box, though *Ilex crenata* 'Convexa' is at least as good for hedges of moderate height in such climates. European box and holly are also suited to hot climates.

At Arley, it is not just their present magnificence that makes its borders of

interest but the record of changing attitudes to essential considerations of scale, colour and texture, highlighted in Victorian paintings of the borders and in the memories of Rowland's great-granddaughter, Lady Ashbrook, of different styles of planting from the 1920s to the present day.

The earliest surviving images of the borders – an 1889 watercolour by Piers Egerton-Warburton and paintings by George Samuel Elgood in the same year – show small groups of herbaceous plants and annuals, a jumbled effect in which the scale is small and the attention to form and foliage texture minimal. There are hollyhocks, red hot pokers, gladioli, phlox, Japanese anemones, sunflowers and sweet peas, while nasturtiums jostle cheekily in front. In spite of Loudon, Victorian gardeners had come to love the combination of gloriously informal planting in a formal framework as much as we do today. An account in *Country Life* in 1901 mentions 'larkspurs and foxgloves, poppies and phloxes, snapdragons and lupins', plants that would seem equally at home in the most humble cottage garden. Lady Ashbrook considers that 'the prestigious thing then was to have every bit of the border filled with colour; there was no nonsense about foliage or form'.

So by the beginning of the twentieth century there was no sign of the regimented planting of Loudon or earlier gardeners, and planting was designed to produce a riot, rather than to supply individual blooms of perfection and beauty, to be appreciated singly at close quarters. But the scale of planting remained small. This spottiness was swept aside by the Edwardians who started to group plants on a much larger scale. The worst excesses of the Edwardians still linger in Lady Ashbrook's earliest memories of the Arley borders. Her mother, like most ladies of her day, 'would have called herself a gardener' but 'really knew absolutely nothing'. Lady Ashbrook feels that 'clumps need to be a *reasonable* size: my mother wanted them to be

enormous'; the Edwardian age was 'the most dreadful period of gardening; the vulgarity was horrendous, all show and ostentation'.

The Edwardians had their prejudices. For instance, at Arley white flowers were disliked, regarded as too reminiscent of funerals, though white phlox were allowed. Lady Ashbrook considers that 'the biggest change in my lifetime is the fact that now white flowers are appreciated. But you must have good whites, the things that have a real sparkle.' This might equally be said of silver foliage, which has a similar leavening effect. Repeating the same item at regular intervals, particularly plants with a strong presence such as onopordums, gives shape and rhythm.

The teachings of the prophet Jekyll, though widely known among early twentieth-century theorists of garden design, had yet to filter through to chatelaines of country houses responsible for directing their head gardeners in the planting of leading gardens of the day: Lady Ashbrook's early awareness

of the advantages of careful use of colour and texture applied to the Arley borders were the result of her own observation rather than a study of Miss Jekyll.

Most gardeners now, planting for effect, will place each plant so that it provides an agreeable harmony or contrast with each of its neighbours. The spottiness of Victorian grouping, in spite of its considerable charm, is clearly not ideal for this: the number of plants in each eyeful is too many for the separate counterpoints between adjacent plants to be appreciated to the full and any effective combinations are to some extent obscured by a muddle of other plants. Nor was the Edwardian move to far larger groups the remedy because it is only that part of the group which reacts with its neighbours that functions fully in the border: if the centre is so far removed from neighbouring groups that it becomes redundant reiteration, although it fills space, it cannot add to the pleasurable experiences of plant association.

ABOVE *At Arley, hedges and buttresses show the architectural value of yew topiary. Allowing hedges to grow wide and tall can destroy garden proportions and make extra work: though few would want to change this splendidly sculptural yew at Montacute, Somerset* (TOP LEFT), *another hedge there* (BOTTOM LEFT) *has had its top removed and sunny side cut back to the main stems; the shadier side will be pruned within three years when the rest has regrown. The former spread is shown by the re-sown grass in front.*

OVERLEAF *The plant associations in the Arley borders stress the importance of contrast among shapes and textures, not just of leaves but also of flowers and even seed pods. Burning* Crocosmia *'Lucifer'* (LEFT) *is thrown into startling prominence by a background of soft blue* Campanula lactiflora, *the more subtle contrasts of echinops, macleaya and ceanothus providing a restful backdrop. A froth of lilac goat's-rue* (CENTRE) *counterpoints the stiff spikes and pointed leaves of ice-cool* Lysimachia ephemera. *Flats, globes and spikes mingle where glaucous poppy heads repeat the shape of* Allium giganteum (RIGHT).

The beauty of individual plants, enjoyed in a good plant collection, botanic garden or even a nursery, is greatly exceeded by that of well-judged combinations of two or more plants.

Borders on the grand scale are often viewed from further away than modest ones and some increase in the scale of the plants and the size of the groups is necessary; but if the distance between viewer and border is so great that the individual shapes and textures of flowers and foliage disappear then the design is not working as it should. Either bolder plants are needed or the dimensions need to be reduced. Such perils become apparent in the grand double borders at the Royal Horticultural Society garden at Wisley in Surrey: walking alongside one border, the viewer gets tired of a long group of a fine-textured plant before reaching the end; looking at the opposite border, 11m/12yd away the scale of the group

is acceptable but the plant appears as an amorphous fuzz. In such grand schemes, there is a risk that the benefits of an increase in scale are outweighed by the losses, plant form becoming too small to register.

So perhaps it is not entirely the prejudices of today that allow us the smug satisfaction of considering the usual scale of contemporary planting – just big enough to accommodate in an eyeful the interplay of two or three adjoining groups – to be ideal. The evolution of the Arley borders, from Victorian clutter, through Edwardian excess to the more balanced scale of today, demonstrates trends found in many, if not most, gardens during the last one hundred years.

There are subtle differences between the two sides: the cooler north-facing border is less light and considerably more damp than the sunny side, where the brick wall behind

rules out certain colours that would be perfectly acceptable against rich green yew. The more favourable microclimate of the warmer side also gives less hardy subjects a better chance of survival.

Some of the plants are old favourites that have grown here since Lady Ashbrook's childhood; these include astrantias (*AA. major*, *m. rubra* and *maxima*), *Sidalcea candida*, solidago, three of the surviving sorts of phlox, heleniums, common montbretia and galega. Lady Ashbrook enjoys the dusky tones and graceful habit of the annual purple orach, though she feels it needs lightening with paler flowers or foliage; this seeds itself obligingly throughout the borders and is allowed to stay where it will contribute to the effect, giving rhythm and repetition in much the same way as the onopordums. Lady Ashbrook feels that shrubs here do not look right: only the dwarf

The impact of planting at close range:

OPPOSITE, LEFT *White achillea and the silver foliage of* Artemisia ludoviciana *provide a foil for bold spikes of a summer-flowering kniphofia, its repeated verticals and dagger leaves harmonizing with verdant crocosmia.*

OPPOSITE, RIGHT *In a harmonious setting of maroon berberis and royal purple clematis, a cloud of galega is enlivened by glowing cerise phlox and ruffled poppy leaves.*

ABOVE, LEFT *The colours of creamy yellow* Anthemis tinctoria *'E. C. Buxton', blue campanula and stark white phlox are each indispensable to the others.*

ABOVE, RIGHT *A delightful jumble of alstroemeria, anthemis, Shirley poppies,* Lychnis coronaria *and seldom grown* Lysimachia ciliata *makes a riotous cottage-garden hotch-potch of glowing colour.*

purple berberis is allowed. Conversely, she feels that plants such as old-fashioned pinks and pansies are so enchanting that they are essential. 'Spikes' (hollyhocks, verbascums and delphiniums) are contrasted with 'flats', particularly the achilleas.

Tommy Acton remembers far more annuals being used in the past, phacelia, *Salvia patens* and plants he feels we dismiss as being 'almost rubbish' such as pot marigolds and larkspur. He has continued to add a few new perennials each year, the best one or two being allowed to oust the less successful older plants; several are selections from seed-raised mixtures such as the various colour forms of *Campanula lactiflora*. Particularly important among later additions have been macleaya, white galega, *Allium giganteum* and the splendid crocosmia hybrids raised by Alan Bloom such as 'Lucifer', glowing infernally against a background of

heavenly blue campanula and echinops.

Some promising new additions that have yet to prove their worth are *Achillea* 'Anthea', another Bloom introduction in soft yellow, not so bright as 'Moonshine' but perhaps less insistent on regular division if it is to keep flowering, and *Dahlia* 'Tally Ho', a variety that shares the divided purple leaves and red flowers of 'Bishop of Llandaff'. It remains to be seen whether this is an improvement but it is the constant search for plants even better than the old favourites that keeps the borders alive, refreshed by a yearly input of new plants.

The borders are cut down after the garden closes in autumn, though they are sometimes left for a few weeks if the frost spares the dahlias. Replanting is generally carried out every five years, sometimes when stock is required for the nursery run here by Lady Ashbrook's daughter, Jane Foster, whose husband Charles is the estate's historian and

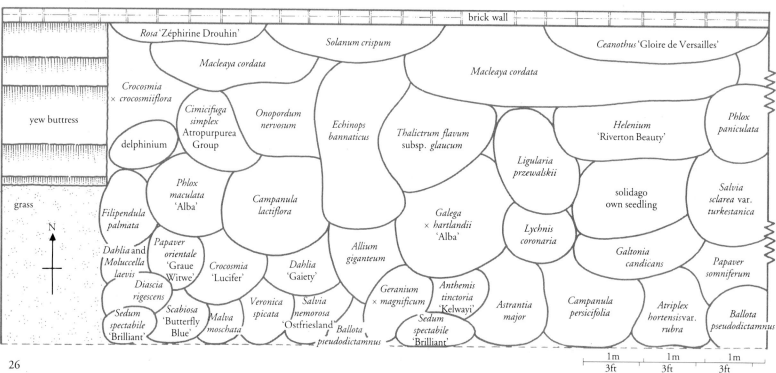

brick wall

Rosa 'Zéphirine Drouhin'

Solanum crispum

Ceanothus 'Gloire de Versailles'

Macleaya cordata

Macleaya cordata

Crocosmia × crocosmiiflora

yew buttress

Cimicifuga simplex Atropurpurea Group

Onopordum nervosum

Echinops bannaticus

Thalictrum flavum subsp. *glaucum*

Helenium 'Riverton Beauty'

Phlox paniculata

delphinium

Ligularia przewalskii

Salvia sclarea var. *turkestanica*

Phlox maculata 'Alba'

Campanula lactiflora

Galega × hartlandii 'Alba'

solidago own seedling

grass

Filipendula palmata

Lychnis coronaria

Galtonia candicans

N

Dahlia and *Moluccella laevis*

Papaver orientale 'Graue Witwe'

Crocosmia 'Lucifer'

Dahlia 'Gaiety'

Allium giganteum

Papaver somniferum

Diascia rigescens

Veronica spicata

Salvia nemorosa 'Ostfriesland'

Geranium × magnificum

Anthemis tinctoria 'Kelwayi'

Campanula persicifolia

Atriplex hortensis var. *rubra*

Sedum spectabile 'Brilliant'

Scabiosa 'Butterfly Blue'

Malva moschata

Ballota pseudodictamnus

Sedum spectabile 'Brilliant'

Astrantia major

Ballota pseudodictamnus

1m 3ft 1m 3ft 1m 3ft

26

archivist while her brother Michael Flower is responsible for its running. They will provide the continuity every garden needs if it is to survive, ensuring that the best of the essence of Arley is preserved while reworking and renewal of the planting keeps it fresh.

Miss Jekyll, writing in *Some English Gardens* in 1904, recognized Arley's essential character, a satisfying blend of formality and natural exuberance: 'Throughout the length and breadth of England it would be hard to find borders of hardy flowers handsomer or in any way better done than those at Arley.... Here we see the spirit of the old Italian gardening, in no way slavishly imitated but wholesomely assimilated and sanely interpreted to fit the needs of the best kind of English garden of the formal type, as to its general plan and structure. It is easy to see ... how happily united are formality and freedom; the former in the garden's comfortable walls of living greenery with their own appropriate ornaments, and the latter in the grandly grown borders of hardy flowers.'

Though the planting has since been much improved, this essential part of the character of the garden remains unchanged, a tribute to sensitive conservation by generations of a single family.

LEFT, TOP *The yew hedge with its distinctive finial, spiky crocosmia and candelabra of onopordum bestow an architectural framework for a tapestry of flowers.*

LEFT, BOTTOM *The borders depend not only on new varieties and sorts raised from seed in the garden but old cultivars such as* Phlox maculata *'Alba' and* Helenium *'Riverton Beauty', difficult to obtain now in Britain but unsurpassed for achieving the required effect. Groups vary in size from 5m/17ft for large plants such as macleaya at the back to small clumps in front. Annuals like poppies and orach sow themselves throughout; though their spread cannot be shown on plan, they are an important element of the design.*

RIGHT *Repetition of simple shapes – heads of poppy and astrantia, plumes of macleaya and wisps of orach – woven through each other gives an interplay of form and texture that is successful without bright colour.*

A BORDER IN THE 'MINGLED' STYLE

Packwood House

The essence of the borders at Packwood House in Warwickshire is the cramming together of an astonishing abundance of flowers, jostling in a space which logic demands should hold only half the quantity. The garden is a fascinating survival, its layout of courtyards, terraces, gazebos and mount remaining intact from the sixteenth and seventeenth centuries when the house was built, but the border planting is typically early nineteenth century in appearance, following Loudon's 'mingled' style in its use of small groups or single plants repeated at intervals with flowers in riotously mixed colours. However, Loudon's dictum that plants in a formal border must be spaced and separated on a regular grid is ignored.

Though early Victorian in style, the planting at Packwood is not a direct descendant of that age but has evolved mainly during the twentieth century through the influence of successive owners and gardeners. Baron Ash, owner and restorer of Packwood who donated house and garden to the National Trust in 1941, created most of the borders in their present positions and probably set the parameters for planting, though they have developed considerably since his day.

The main border, traditionally known as the Yellow Border though that colour scarcely predominates, faces the sun and is backed by a hollow brick wall, once heated to protect the delicate blossoms of peaches against spring frosts. With planting as high as the border's

The Yellow Border's combination of pink, yellow, lavender, carmine and scarlet flowers against russet-orange brick might shock those of a nervous disposition. Perhaps these recent years of colour scheming have made us all too sensitive: after all, this is just the sort of joyous mixture we love to see in a cottage garden or flower meadow, brought to its cultural apogee by the gardener's skill. Fortunately, the pink of Filipendula rubra *'Venusta' and the*

lemon of Potentilla recta *var.* pallida *are just gentle enough to avoid the worst of clashes and sprays of plume poppy* (Macleaya cordata) *help isolate the candyfloss pinks from the brickwork.*

29

ABOVE *The Yellow Border lies at the far side of the Sunken Garden, where purple flowers traditionally predominate. The Sunken Garden is of a type common in early twentieth-century England, often called a Dutch garden and supposedly a nostalgic recreation of a style dating back to the time of William and Mary. Similarities with the Pond Garden at Hampton Court, often assumed to date from William and Mary's reign though actually later, are apparent.*

The arch beneath the gazebo steps housed the fire for heating the wall, behind which lies the East Court with its Blue Border. To the right of the gazebo are hedged rose beds. Packwood's most famous feature is the Yew Garden (see plan RIGHT*), separated from the main flower garden by a raised terrace walk. Said to represent the Sermon on the Mount, it is mainly Victorian, only the yews at the farthest end being older. Its comparatively recent origins have not stopped it from being held up as an archetype of a Renaissance formal topiary garden.*

OPPOSITE *Such a tall and narrow border with so many plants crammed more closely than usual would be impossible to achieve without support. Staking is prepared in spring. The brushwood forms a single sloping plane, analagous to the netting used at Cliveden, holding each stem firm at a little below its midpoint. This method is not ideal for skyscraper plants such as delphiniums which prefer higher support; only the sturdiness of their stems, through excellent cultivation, keeps them upright. To the gardeners' credit not a twig of the thicket of brushwood that holds up this border is visible in summer. As spring ends, wallflowers are replaced by tender perennials such as argyranthemums and dahlias.*

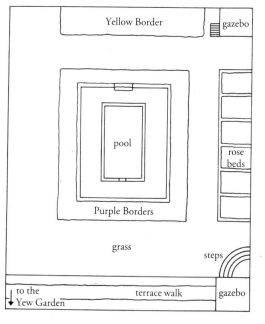

width (2.4m/8ft), growing at the back almost to the full height of the wall behind, the border seems full to overflowing, bursting with flowers of every hue from top to toe. Such steeply sloping borders demand plants well furnished with flowers for a considerable proportion of their height; flat-topped plants such as some of the achilleas, which would appear as a line of colour with an expanse of green beneath, are less successful for such towering planting and are not used here.

Some of the traditional rules of border making have been defiantly jettisoned: that the tallest plants should have a height of about half the width of the border and that each group should be spaced sufficiently far apart to ensure sturdy growth and minimize staking. Such a crowded and steeply banked effect becomes possible only if the border is staked thoroughly from one end to the other. Because groups are small and there are no gaps between them, individual staking is not an option. The border must grow up each year through a continuous matrix of stakes which holds every plant in its rightful place.

The task of staking begins in the winter months when Head Gardener Fred Corrin and his staff gather some forty bundles, each of about twenty-five tall peasticks, from the nearby woods. This large quantity is sufficient for the two main borders (the Yellow Border and, in the East Court, the Blue Border) with some sticks reserved for odd groups of plants elsewhere in the garden. The staking is tackled in mid spring, when plants are no more than 15cm/6in high, starting behind the front rank of plants which are low enough not to need support. The brushwood is angled towards the wall at about forty-five degrees, reaching a height of about 1.35m/4½ft at the back, thus accommodating all the plants in the gradient from the shortest to the tallest. Though many gardeners would not like to see so many stakes, there is something satisfactory about the craftsmanship evident in the neat bank of brushwood, which anyway disappears among the growing plants within six weeks or so.

The requirements for sloping stakes are very different from the more usual vertical ones: for skyscraper plants like delphiniums, which must be able to sway if they are not to snap, upright supports should have a degree of flexibility; widely divergent branches which would poke out of the side of the group and would be difficult to work in among part-grown plants are not an asset. At Packwood, the brushwood must hold everything in its place with a vice-like grip; birch and lime would be too pliable and a mixture of elm, oak, sycamore and hazel is preferred. Though not much use for vertical stakes, the more markedly branched sorts are important for locking the whole together into a rigid structure. Poplar and willow are avoided because even such large sticks would root and grow. Though a path along the back of the border would be useful for tending wall plants, space is deemed to be too precious to spare here and roses and other climbers must remain untended while the brushwood is in place.

The carefully maintained balance between so many small plant groups and the obvious vigour of the plants might seem to be the result of yearly replanting of a large proportion of the border's contents. This is not the case, though the gardeners are very careful to note which plants have increased and which have dwindled as the season progresses. When the border is worked through in spring, overlarge groups are reduced and any that are too small are bulked up. *Aster tradescantii* is the worst offender among the spreaders and needs to be curbed drastically every year. Delphiniums regularly prove unequal to the competition, particularly if they have attracted the attentions of an errant slug, but are so important for their stature and the richness of their colour that they are always replaced. *Salvia sclarea* var. *turkestanica* is the only biennial in the main body of the border, raised from seed each year and replanted the following spring.

The contents of the border do not change greatly from year to year, though occasionally a new plant is introduced to give longer or more abundant flower. Although there is no attempt to use historic plants, all the plants

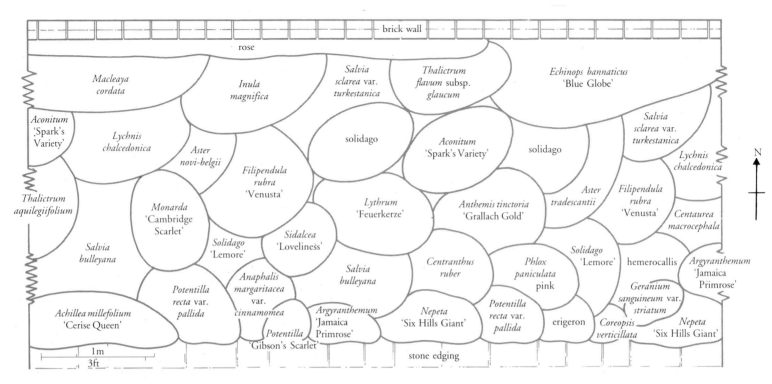

brick wall

rose

Macleaya cordata

Inula magnifica

Salvia sclarea var. *turkestanica*

Thalictrum flavum subsp. *glaucum*

Echinops bannaticus 'Blue Globe'

Aconitum 'Spark's Variety'

Lychnis chalcedonica

Aster novi-belgii

solidago

Aconitum 'Spark's Variety'

solidago

Salvia sclarea var. *turkestanica*

Filipendula rubra 'Venusta'

Lychnis chalcedonica

Thalictrum aquilegiifolium

Monarda 'Cambridge Scarlet'

Lythrum 'Feuerkerze'

Anthemis tinctoria 'Grallach Gold'

Aster tradescantii

Filipendula rubra 'Venusta'

Centaurea macrocephala

Sidalcea 'Loveliness'

Solidago 'Lemore'

N

Salvia bulleyana

Solidago 'Lemore'

Centranthus ruber

Phlox paniculata pink

Solidago 'Lemore'

hemerocallis

Argyranthemum 'Jamaica Primrose'

Potentilla recta var. *pallida*

Anaphalis margaritacea var. *cinnamomea*

Salvia bulleyana

Geranium sanguineum var. *striatum*

Achillea millefolium 'Cerise Queen'

Argyranthemum 'Jamaica Primrose'

Nepeta 'Six Hills Giant'

Potentilla recta var. *pallida*

erigeron

Coreopsis verticillata

Nepeta 'Six Hills Giant'

Potentilla 'Gibson's Scarlet'

stone edging

1m
3ft

ABOVE *This section of the border, seen opposite in midsummer, represents one block of a repeated pattern, the larger, taller plants at the back recurring less frequently than the shorter plants in front. However, though each plant reappears at more or less regular intervals, the order and pattern are slightly altered to avoid monotony and create different plant associations. The stone margin lets the plants sprawl forwards, softening the edge of the border.*

LEFT *At the start of the border's early summer display, the brushwood is already enveloped by the plants. Repeated groups of* Nepeta *'Six Hills Giant' and* Centranthus ruber *towards the front give a strong rhythm. Lavender and carmine-red flowers predominate, with no white or orange and little yellow.*

RIGHT *In midsummer flowers of almost every hue jostle for attention.* Monarda *'Cambridge Scarlet',* Aconitum *'Spark's Variety' and* Anthemis *'Grallach Gold' dominate; the catmint and red valerian are more subdued as they reach the end of their season.* Salvia bulleyana *adds another quiet note, its primrose flowers long in bloom, charming but never brash.*

must look like traditional border flowers, even if they happen to be choice new selections. The gardeners are aware of the pitfalls of pandering to 'good taste', following current garden fashion, or even of copying Gertrude Jekyll, sure ways of losing the individuality of the garden. The only style which is followed is that of Packwood itself.

Some old favourites are to be found in the Yellow Border, many of them more typical of early twentieth-century gardens than modern ones, such as heleniums, goldenrod, echinops, anthemis, phlox and sidalcea; some – for instance *Helenium* 'Riverton Beauty' with brown-eyed flowers of the most rich and vibrant yellow – have long since disappeared from British nursery catalogues though they remain superlative for borders of this sort.

Fred Corrin feels that visitors enjoy the unrestrained mix of flowers, reminiscent of a cottage garden writ large. Throughout the summer, the tiniest patch of bare soil is regarded as the most reprehensible of faults. The gardeners aim to provide a riot of colour so that during each of the summer months new varieties will come into bloom for every sort that fades.

The structure of the planting depends much less on contrasts of habit or foliage texture used so successfully by Gertrude Jekyll and her followers. At Packwood it is the rhythm of repeated plants, generally about four or five groups of each sort, and the colours of their flowers that give pattern and emphasis. In early summer, reds and blues predominate with scarcely a hint of yellow to be seen. Delphiniums and anchusas add height with pinkish-red valerian through the heart of the border and a haze of lavender catmint 'Six Hills Giant' towards the front sprawling across the stone edging. At 30cm/12in wide, the stones allow each group at the front of the border exactly the right space in which to flop without encroaching on the lawn.

Before midsummer is reached, the yellows appear: in the heart of the border, a primrose

form of *Salvia bulleyana*, without the brown or violet lip sometimes found in this species, is one of the border's more unusual plants; groups of *Anthemis* 'Grallach Gold' provide a much more dominant note between each group of salvias, while towards the front *Potentilla recta* var. *pallida* and *Argyranthemum* 'Jamaica Primrose' drape themselves across the stone edging between the catmints. At this time, the queen of the prairie, *Filipendula rubra* 'Venusta', begins to flaunt fluffs of candyfloss pink in front of an ivory haze of the majestic plume poppy, *Macleaya cordata*.

This is perhaps the moment of the border's greatest glory and its most exuberant mixture of colour. When midsummer arrives, the pink of the filipendula begins to darken to russet red, adding a prematurely autumnal tone. There are still plenty of flowers that remain in bloom throughout the rest of the summer, though the appealing freshness of those early months is lacking.

After the garden closes to the public in autumn, the brushwood is removed along with the haulm of all the herbaceous plants, leaving just enough stem to show the position and identity of each. Then the gardeners are able to spread farmyard manure, thoroughly rotted for as much as two years, over the borders, taking care not to cover the crowns of the plants. After the borders are staked in spring, wallflowers provide an early display along the front in spaces between permanent plants such as potentillas and catmint; when they are removed in late spring or early summer their place is taken by long-flowering tender plants such as argyranthemums and short pompon dahlias. A little rose fertilizer or bone meal is the only other feed the borders receive, the manure providing an ample supply of nitrogen.

The thorough staking gives little opportunity for weeding but fortunately the closeness of the plants prevents all weeds except goosegrass from competing. Through

LEFT *A contrast of red and yellow teams well with a cheerful medley of different forms. The slightly jarring reds of* Monarda *'Cambridge Scarlet' and* Achillea millefolium *'Cerise Queen' give a tingle of discord while yellow* Potentilla recta *var.* pallida, Anthemis *'Grallach Gold',* Salvia bulleyana *and* Centaurea macrocephala *are altogether more harmonious.*

RIGHT Argyranthemum *'Jamaica Primrose', planted next to* Nepeta *'Six Hills Giant' for contrast, is one of few tender perennials used at the front of the border to fill the gaps left when wallflowers are removed.* Potentilla *'Gibson's Scarlet' weaves through the front of the border, its bright flowers cheekily interrupting the gentler tones.*

the summer, the main task is to hose with water from the garden's lake; it is perhaps the fact that the plants never want for water that makes them all so luxuriant: *Lychnis chalcedonica* reaches 1.8m/6ft high and macleaya an astonishing 2.4m/8ft. Some spraying is necessary to prevent blackspot and mildew on roses along the back wall. The only other pesticide used is an occasional scattering of slug pellets, easily applied through the forest of brushwood.

The style of planting at Packwood is charming in its unsophisticated simplicity and a triumph of the gardener's skill in its astonishing abundance. Visitors who are not aware of the elaborate staking beneath may well wonder how so much tightly packed and luxuriant growth manages to stay up at all. As Fred Corrin points out, there are more 'tasteful' schemes in plenty of gardens but gardeners are still pleased to see the Packwood borders, refreshingly different and a unique apotheosis of the cottage garden style.

DOUBLE SUMMER BORDERS

Nymans

The Summer Borders at Nymans, at Handcross in Sussex, are abundant, overflowing and exuberant, full of old-fashioned flowers. Gorgeous dahlias, free-flowering annuals and bold cannas are backed with tall perennials in a display of *fin de siècle* magnificence, reminiscent of the paintings of George Samuel Elgood or Beatrice Parsons. Their character survives unchanged, perhaps because there have been only three head gardeners at Nymans since the borders were made: James Comber who first planted them, Head Gardener from the turn of the century until his retirement aged 86 in 1953, just before Nymans was bequeathed to the National Trust; Cecil Nice, who kept them unaltered for almost thirty years; and David Masters, Head Gardener since 1980. Though

there is no attempt at sophisticated colour scheming, the most brash of modern varieties are avoided to give a tapestry of colour, tumbling with flowers from top to toe. The richness, intensity and length of display are due as much to the excellence of horticultural technique and the production of superbly grown plants as to the choice of varieties.

Effectively a double border interrupted at its midpoint by an ornate Italian fountain, there is no record of the making of the borders nor of their designer. Tradition has it that both Gertrude Jekyll and William Robinson had a hand in the design; Robinson certainly visited but Miss Jekyll's name does not appear in the visitors' book. Perhaps their involvement was little more than a stroll through the garden with Ludwig Messel, the

Running along the central axis of the Wall Garden, once the estate's orchard near to the house, the Summer Borders are at their most glorious in late summer, though they continue in bloom until the frosts. Although many of the varieties are modern, none would look out of place in a garden of 1900. In borders like these, narrow relative to the tall plants, evenly graded heights ensure a wall of bloom, each

plant covering the stems of the variety behind. Dark foliage (here Dahlia *'David Howard') is used sparingly, to add richness and to punctuate long stretches of green-leaved varieties, but never to the point of making the effect too heavy. There are no variegated leaves, partly because they are rare among annuals on which the border depends but also because they could make an already busy colour scheme unbearably frantic.*

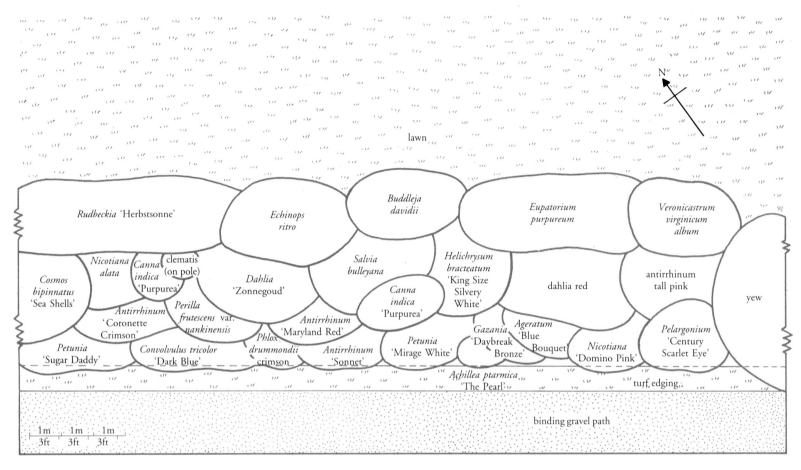

The northwest–southeast alignment of the borders is a successful compromise between a north–south orientation, giving equal light on both sides, and east–west, which gives dramatic lighting in low morning and evening sun. Even the shadier northeast-facing border is bright enough for its planting to match exactly the sunnier border shown here, without any evidence of poor light. Tall, narrow perennials and buddlejas at the back of the border make up for the lack of any hedge or wall behind; their height of 2.5m/8ft gives a feeling of total enclosure. The front three ranks of the border consist of about two-thirds annuals, with cannas, dahlias and a few long-flowering herbaceous plants such as Achillea ptarmica *'The Pearl' and* Salvia bulleyana *making up the remainder. The borders' relatively narrow width and steeply ascending heights make it difficult to use groups which slant away from the path; almost all the groups here run parallel to the front of the border.*

garden's creator, and a few encouraging and complimentary comments.

Annuals account for more than half of the varieties in the Summer Borders. Like tender perennials, they are valuable for flowering late in the summer and on into autumn when the range of long-flowering herbaceous plants is extremely limited. Every autumn, David Masters, his Assistant Head Gardener Philip Holmes and Rob Massen who raises the seedlings, scour the seed catalogues and consider any changes to the range of annuals. Only a handful of varieties is altered each year, most of these simply improvements on old favourites, longer flowering and more disease- or weather-resistant.

Any additions must remain absolutely in character with the traditional planting and the brightest and brashest colours, brilliant

pink, yellow or orange, are avoided. French or African marigolds are felt to be too strident and some pink petunias have proved too vivid. The plantsman's temptation of including many small groups of yet more varieties, destroying the scale and making the planting excessively complex, has to be resisted. David Masters feels that the predominance of annuals is a feature of the border that must remain; tender perennials, in spite of their current popularity, are little used here, although they feature in several other parts of the garden.

Some recent additions have proved to be both successful and in character and have been added to the borders' regular repertoire: cleomes, *Helichrysum bracteatum* and heliotropes have done well even in the wettest and most dismal summers. Californian

poppies were not a success, lasting in flower for barely a month and leaving a gap for the rest of the season. Shirley poppies also finish blooming well before the borders' other flowers but are felt to earn their keep.

The varieties of annuals to be sown are listed, numbers required of each determined and a sowing programme is drawn up. Seed is sown over seven weeks, starting in late winter (usually the fourth week in February). This ensures that all annuals reach planting size at the same time in late spring after the danger of frosts. A peat-based compost containing a little sand is used for sowing. Peat alternatives were found to need a much more precise watering regime than the gardeners, with a further 12 hectares/30 acres of garden to maintain, could provide. Fine seed is sown onto a thin layer of vermiculite, which

prevents the compost from forming an impermeable cap; its pale colour also prevents overheating in sun. The seed trays are then lightly watered to take the seed into the surface of the compost. Larger seed such as cosmos, China aster or *Salvia coccinea* is covered with a layer of vermiculite as thick as the seed.

Seedlings are pricked out twenty-four to a tray into peat-based potting compost to which slow-release fertilizer, fungicide (furalaxyl or etridiazole) and silicone-based water-retentive polymer have been added. Larger-growing plants are pricked out into 87mm/3°in pots. Such generous spacing and pot sizes ensure vigorous bushy plants; fewer will be needed and all will last much longer in flower than plants grown more closely together as seedlings. Çleanliness in every stage of

In the section of border shown on plan, abundant bloom is the prime consideration, but cannas are included for the bold architecture of their leaves. Modern hybrids may have showier flowers but it is hard to surpass Canna indica *'Purpurea' for its statuesque habit, effective as a repeated accent in all four sections of border, especially when it is placed slightly in front of plants of similar height. Purple-leaved* Perilla frutescens *var.* nankinensis *is also grown for foliage effect and the feathery leaves of cosmos are almost as pretty as its blooms. Australian strawflowers (*Helichrysum bracteatum, *on the right) are seldom used in the flower garden, perhaps because of their untidy foliage; here their leaves are hidden by the surrounding plants. Even in the wettest summers these everlasting flowers have proved showy and reliable.*

propagation is crucial; all pots and trays are thoroughly scrubbed and washed with fungicide to minimize any risk of damping-off diseases such as rhizoctonia, a threat to most of the annuals but particularly to antirrhinums and lavateras.

The compost into which the seedlings are transplanted must have the right moisture content, just damp enough so that it holds together when squeezed without water dripping out. Watering thereafter is critical. After transplanting, the seedlings are lightly sprayed over, giving them rather more water when they show signs of establishment and the beginnings of growth. In these early stages, the tiny seedlings are unable to cope with surplus water in a sodden compost and would rot. The temperature regime needs to be constant; at Nymans, a daytime temperature of 10–12°C/50–55°F is preferred, rather lower at night. Higher temperatures can result in etiolated and sappy plants which suffer a severe check when planted out.

Biological control is used to combat glasshouse pests. Although the temperatures are rather low for this, the results have been very satisfactory against vine weevil, red spider, whitefly and aphids. Timing is crucial: at the first sign of attack, the predators must be ordered so that they can be introduced to the glasshouse before the pests have multiplied out of control. Late infestations of aphids can be treated using horticultural soap, though this cannot be used on the youngest seedlings.

As planting time approaches, the borders

must be prepared. The perennials that form the backing to the borders must be staked while the ground in front is still empty; the closely packed annuals would make it difficult to work through the border later on. Some 3000 hazel peasticks, half of them 2.4m/8ft, the other half 1.5m/5ft, are bought in for the job. Though unsightly for a time, the brushwood soon disappears completely as it is enveloped by the perennials. The hazel stems are worked first into the middle of each group to prevent the centre of the groups falling out towards the edge, then placed through the perimeter of the clump; the tops of the stakes are then broken over.

An inorganic fertilizer with high potash content to encourage flowering is applied to the border before the dahlias are planted in late spring (usually the second week in May). Then the border is marked out with sand where the annuals are to go. The gardeners walk down the borders with the list of plants and decide on the positions of each variety from one end of the borders to the other. Heights are evenly graded towards the back and textures varied for contrast. Though adjacent groups of the same colour are avoided, there is no colour scheme: some of the colours clash, others blend together harmoniously. Generally, ten trays of each variety are grown, two per half border plus two spare, so that one group of each variety is used in each of the four quarters. The size of each group has remained the same for many years and is regarded as the optimum for this sort of planting. Bigger groups would result in excessively large gaps where any variety passes out of flower, destroying the visual continuity and the interplay of colours along the border.

Annuals that need staking, such as cleomes, lavateras or antirrhinums, are tied to bamboos or split canes with plant rings. Spraying against pests and diseases is rarely necessary, though occasional attacks of thrips or blackfly are dealt with as they arise. Some

deadheading is needed for the dahlias and especially to make the antirrhinums produce a second flush of flowers. Weeding and possibly irrigation are the only routine maintenance jobs through the summer months.

Though the healthful appearance of the young plants is appealing and full of promise, it is not until midsummer that the display begins in earnest. By then, the bare poles that support the clematis have become decently clad, swagged and studded with bold flowers in purple, blue and white. In late summer, the dahlias come into their own; the maroon foliage of perillas and cannas has assumed sufficient bulk to add gravitas and richness, though David Masters feels this colour should only be used in moderation. Inevitably some annuals are past their peak by this time and have lost some of their freshness. David Masters is considering replacing them with chrysanthemums or perhaps late sowings of yet more annuals so that the borders in autumn lack none of the brilliance of the summer display.

As soon as the garden closes in mid autumn, the annuals are stripped out of the borders. Dahlias are also removed, even though they have not usually then been frosted; David Masters feels there is no benefit in waiting until they have become black and slimy, as some gardeners believe. They are then put upside down for a week or so to dry, letting water drain from the bottom of the stems and making easier the removal of the soil. The roots are then stored in boxes under the potting shed bench with no covering.

Keeping the tubers naked allows easy inspection so that any rot or excessive desiccation can be rapidly tackled. If they start to look shrivelled, they can be sprayed lightly with water and will quickly become plump again. Early in the spring, when the glasshouses are damped down, the roots will again be sprayed with water; by the time they are planted out, about 10cm/4in of top growth has appeared. Avoiding growing them

ABOVE *When the annuals are newly planted, there is little bloom to see in the Wall Garden save for the magnificent dogwood,* Cornus kousa. *The unsightly brushwood stakes and bamboo canes supporting young cosmos will have vanished by the time the display starts in earnest in midsummer. Annuals are planted a short distance away from the turf verge; by the time they are in full flower, the bare earth will have been covered.*

OPPOSITE *Well sited to glow in the rays of the setting sun, the borders succeed without a backing of hedges to give formality, and the walls, seen at the end of the border, are so thoroughly clothed that their architecture plays little part in the garden's appearance. Structure is provided by the elaborate yew finials, fountain, arch and formal turf edgings underlining the borders' boisterous colour. Each of these elements is as essential to the success of the garden as the flowers themselves.*

in compost saves a lot of work and no tubers have ever been lost. If more plants are needed, the tubers are divided; it is not generally necessary to take cuttings.

Dahlias and cannas are the only tender perennials in the Summer Borders that need to be overwintered under glass. The cannas are potted up when the border is stripped in autumn, dividing each clump into pieces to fill 225mm/9in pots. Keeping them fairly dry at a minimum temperature of 5°C/40°F holds them back until spring when they can quickly be started into vigorous growth.

After the annuals have been removed, the borders are forked through, incorporating mushroom compost in some but not all years: the soil must not become too rich in nitrogen if flowers rather than foliage are to predominate. The slightly alkaline mushroom manure has not altered the pH of the soil significantly and has had no detrimental effect on the range of plants grown. In one winter in three, the herbaceous plants are lifted and divided. All are carefully labelled and removed under cover where they can be split and cleaned of perennial weed roots such as bindweed when inclement weather prevents outdoor work.

Until a few years ago, the perennials included *Phlox paniculata* cultivars until infestation by stem eelworm made them difficult to grow. David Masters hopes to reintroduce them now that the three years that eelworms remain in the soil have elapsed. Some old favourites seldom used today still remain, such as *Aster novi-belgii* 'Climax', *Helianthus salicifolius* and *Silphium perfoliatum*, most of them tall and narrow to provide a dramatically banked effect and give an impression of enclosure without the need for a hedge.

Of all the borders in this book, the Summer Borders at Nymans have changed the least over the decades. While further evolution has benefitted other borders by

even the most eminent designers, the style of the Nymans borders was already fully evolved almost a century ago. For such an eminent example, a rare survival of an historic, romantic and attractive style, conservation is more important than further change. This is not to underestimate the skill and creativity of the Nymans gardeners. To conserve something so naturally changeable as a border without any significant alteration in its character and yet to keep the planting constantly alive, refreshed and renewed from year to year, demands great artistry and technical ability. The Nymans staff possess these qualities in plentiful measure.

LEFT *In any randomly chosen section of the borders, there is an equal chance of finding gentle harmony or vibrant contrast. Here burgundy and carmine dahlias,* Cleome hassleriana *'Rose Queen' and lavender-coloured* Ageratum *'Tall Blue' combine agreeably, their colours lifted by* Helichrysum bracteatum *'King Size Silvery White'. A column of* Clematis *'Ville de Lyon' and lofty* Eupatorium purpureum *provide the necessary height. The lively outline of the carmine Cactus dahlia makes a pleasant contrast in flower shape with that of 'Symbol' behind. The ageratum is one of few taller varieties of this genus, more useful than ground-hugging modern varieties for being able to blend into the contours of bed, border or container.*

RIGHT *Another randomly chosen stretch of border has clashing colours. Because so many of them are late-flowering, the daisy family Compositae accounts for more than one-third of the borders' flowers. Here zinnias, cosmos, red* Dahlia *'Greenside Antonia' and pink* D. *'Rothesay Castle' argue with golden helianthus.* Convolvulus tricolor *'Dark Blue' and* Petunia *'Sugar Daddy' furnish the front. Plants in the Summer Borders are usually arranged to keep different varieties of the same genus separate: the juxtaposition of two dahlias and two helianthus here is an exception.*

RED DOUBLE BORDERS

Hidcote Manor

With the possible exception of purple, red is perhaps the most difficult colour to use effectively in the garden. It has the potential for being at best exciting or at worst irritating. At Hidcote in Gloucestershire, the Red Borders first planted by Colonel Lawrence Johnston are an object lesson in how best to use the colour, making a virtue of its problematical qualities, and are perhaps the single most original and successful feature of the garden. Lying along the main axis of the garden, the Red Borders are processional, a long gallery connecting the other garden rooms, meant to be walked through and not a place to stop and rest. Johnston's choice of flower colour reflects this character, exhilarating, uplifting but far from restful. Originally called the Scarlet Borders, it is clear from early photographs that even then this was a misleading name: since Johnston's day, the flowers have been of every shade of red from vermilion to crimson; to these, Johnston added soft orange daylilies, with rich purple

Aconitum 'Spark's Variety' and maroon foliage to make a dusky foil for vibrant colour.

Graham Thomas has likened the combination of reds to Augustus John's painting of the cellist Madame Suggia whose flowing scarlet dress is dazzlingly portrayed using brushstrokes of exactly the same range of different tones. In a long association with the garden, Graham Thomas has enriched the planting of the borders following Johnston's scheme and applying what he believes to be the cardinal rule about colour combinations: that red containing yellow should not be mixed with red containing blue. Occasionally over the years, plants were added which later proved to disobey this rule such as the Floribunda 'Rosemary Rose', *Buddleja davidii* 'Royal Red' and *Lobelia* 'Dark Crusader', only to be removed once their unsuitability was recognized. Head Gardener Paul Nicholls feels that the combination of all these colours, not just reds but soft orange and dusky purples with a considerable quantity of green

The sunnier of the two borders shows the characteristic mix of plants. Height and dark foliage are provided by Acer platanoides *'Crimson King', now felt by Graham Thomas to be too coarse, black and vigorous.* Lobelia × speciosa *cultivars provide glorious scarlet throughout the middle ranks of the border with foliage contrast from* Miscanthus sinensis *'Gracillimus' and* Hemerocallis fulva

'Flore Pleno'. The doubleness of the daylilies makes their individual blooms last longer; flowering all summer, 'Flore Pleno' has more shapely florets than the similar 'Green Kwanso'. Charmingly informal 'Frensham' roses, phlox and Salvia elegans *fill the foreground, with airy rose 'Geranium' cascading fiery red hips behind. Canna 'Roi Humbert' supplies bold emphasis essential to the exotic richness of the design.*

foliage as well, makes the Hidcote Red Borders more satisfying than others he has seen that use just scarlet flowers with maroon foliage, a recipe he finds as cloying as chocolate fudge cake.

Born in Paris in 1871 to American parents, Lawrence Johnston took British citizenship in 1900 and moved among the well-to-do circle of expatriates sometimes called the 'Henry James Americans', many of them resident nearby in the Cotswolds. Johnston moved to Hidcote with his mother Gertrude Winthrop in 1908; by about 1910 the central part of the garden, including the Red Borders, had been laid out, though the planting had changed considerably by about 1930 when the garden started to become widely known. It is still not clear how Johnston acquired the skill to design a garden of the quality of Hidcote nor who were his main influences. Some have suggested that the structure of the garden owes much to the garden architect Thomas Mawson, and indeed the early photographs of Hidcote in about 1910 show abundant trelliswork of just the sort Mawson favoured; perhaps fortunately, most of this plethora of joinery was soon replaced with hedges. Norah Lindsay, a close friend of Johnston and an eminent garden designer, certainly played a leading role and his many gardening friends and neighbours doubtless provided the inspiration for some of Hidcote's features.

Perhaps from Gertrude Jekyll's books, Johnston had learned the lesson of use of foliage and form; by 1930, fine-leaved *Miscanthus sinensis* 'Gracillimus' and bolder daylilies had been introduced and remain important to the interplay of textures today. But Johnston's most significant innovation was the fusion of the cottage garden tradition with the larger scale of Hidcote: as in cottage gardens, key plants are repeated at intervals along the border, groups of lobelias, begonias or fuchsias being set with odd plants away from the group, an effect that imitates nature, as though seedlings had appeared near the

original plant. Though Miss Jekyll acknowledged her debt to cottage gardening, the Hidcote style of planting was quite different from anything she devised, a distinct advance and probably unique in its day, though now much copied. Paul Nicholls still maintains this tradition of groups with odd plants set away from the group, though he feels it can look contrived if overdone.

Johnston spent many hours working with his own gardeners, supervising rather than dirtying his hands. Staff as expert as Frank Adams, Head Gardener from 1922 until 1939, and Walter Bennett, the quality of whose planting was renowned, were probably as influential in the development of the Hidcote style as many of Johnston's grand gardening friends.

Johnston enjoyed plants for their own sake. If he liked a plant, he would find somewhere to grow it; he was probably not worried if the artistic integrity of his Red

Borders was compromised, a possible explanation for one or two of the odder finds there. Paul Nicholls believes that the large *Pinus mugo* grows in the sunnier of the two borders simply because Johnston wanted to grow it and thought it would thrive there, rather than because he felt it added to the border's appearance; Hidcote was created for his own enjoyment, not as an eternal work of garden art. The pine is certainly the most notable of the borders' eccentricities today, though for all its oddness the borders would be diminished by its loss. Another curiosity, overwhelmed by the ever-spreading yew hedge some years ago, was silvery-blue-leaved *Berberis dictyophylla*, also believed by Nicholls to be an example of the occasional primacy of plantsmanship over artistry.

Harry Burrows, Head Gardener from 1959 to 1978, came to Hidcote while Walter Bennett still worked there, ensuring that the garden's traditions of planting continued

LEFT *Looking towards the house, an old cedar of Lebanon provides a focal point, though slightly off the centre of the garden's axis. Graceful* Prunus spinosa *'Purpurea' gives an ideal foil of dusky foliage while, behind, purple and green tapestry hedges hide all but a tantalizing glimpse of lilacs beyond. Already well-clad with foliage, the borders' colour is supplied by tulips 'Dyanito', 'Red Shine' and 'Queen of Night', occasional rogue bulbs of violet tulip 'Greuze' harmonizing cheekily with the lilacs. Lawrence Johnston's plant of* Pinus mugo *is the borders' one great eccentricity, not active in the colour scheme but adding much character.*

RIGHT, TOP *The bronze-leaved variety of* Cordyline australis *is as important as the cannas in providing punctuation and an air of subtropical luxury, never more so than when glowing against the sun and woven through with* Dahlia *'Bishop of Llandaff'.*

RIGHT, BOTTOM Aconitum *'Spark's Variety', used by Lawrence Johnston to give richness to a dusky background, perfectly sets off the refulgent flowers of dahlias 'Bishop of Llandaff' and 'Bloodstone' in the shadier border.*

uninterrupted. A considerable plantsman responsible for introducing numerous good plants to the garden, Burrows was also renowned for severe pruning, held by some to be Rhadamanthine while others thought it unduly Procrustean. Be that as it may, his secateurs helped keep the balance between trees and shrubs and herbaceous plants from changing as quickly and as drastically as nature would normally dictate. However, the trees that Johnston planted, particularly the holm oaks seen at the far end of the Red Borders and the trees in the Winter Border that now overshadow part of the Red Borders, have grown considerably and utterly changed the scale of the design. It was inevitable that the planting of the Red Borders should change too, becoming more bulky to keep their scale in balance.

Following Johnston's lead of including a purple-leaved Japanese maple, Graham Thomas added a number of shrubs and small trees with dusky foliage to relate the bulk of the borders to that of the trees beyond and to provide a foil for the brilliance of the flowers. Most of these still survive, although even he would admit that not all are equally successful: *Acer platanoides* 'Crimson King' is too vigorous and too near to black to be telling, and both this and *Corylus maxima* 'Purpurea' are rather coarse; the purple plum *Prunus cerasifera* 'Pissardii' is better but still not as graceful and refined as purple sloe, *Prunus spinosa* 'Purpurea', several plants of which cast an airy haze over the borders, their misty greyish tone evidence that the most richly coloured form is not always the best.

Some of the larger plants, particularly buddlejas and miscanthus, are planted towards the front of the borders, varying their contours and preventing an evenly banked effect. However, Paul Nicholls realizes that there is a danger that such plants will block the view of the continued progression of colours along the borders; it is occasionally

necessary to remove or reduce drastically some of these if the interplay of different reds is to remain effective. The larger shrubs and small trees must also be reduced from time to time so that a tantalizing glimpse of at least the outline of the two gazebos is visible along the entire length of this main axis of the garden.

There have been fewer changes among herbaceous plants than among shrubs or tender perennials in the borders. Graham Thomas introduced *Delphinium* Black Knight Group to provide the same colour as *Aconitum* 'Spark's Variety' earlier in the season. *Heuchera micrantha* 'Palace Purple' was introduced to replace *H. americana* which has similar reddish brown leaves but no longer flourished.

Well-rotted manure and compost is dug into the borders after they have been cleared in the autumn but before the tulips are planted. A general fertilizer is also applied when most of the bedding is planted in late spring or early summer. A proportion (usually

about a third) of herbaceous plants such as daylilies and miscanthus is divided and replanted each spring.

Johnston seems not to have minded if parts of the garden had only a short season: in the Pillar Garden, Johnston's borders of peonies bloomed for only ten days or a fortnight. Though we would not want to forfeit altogether such fleetingly beautiful flowers, it is not unreasonable to want a longer display. There is no reason why late borders should not also have something to show in spring. In the Red Borders Graham Thomas added spring flowers including tulips, wallflowers and red Cowichan polyanthus with pulmonarias to cover the ground at the backs and beneath shrubs. The wallflowers and polyanthus are no longer used; many gardeners find wallflowers less reliable than they used to be, perhaps the result of a century of inbreeding, and Cowichan polyanthus are hard to find in sufficient quantity. Tulips 'Queen of Night', 'Ruby

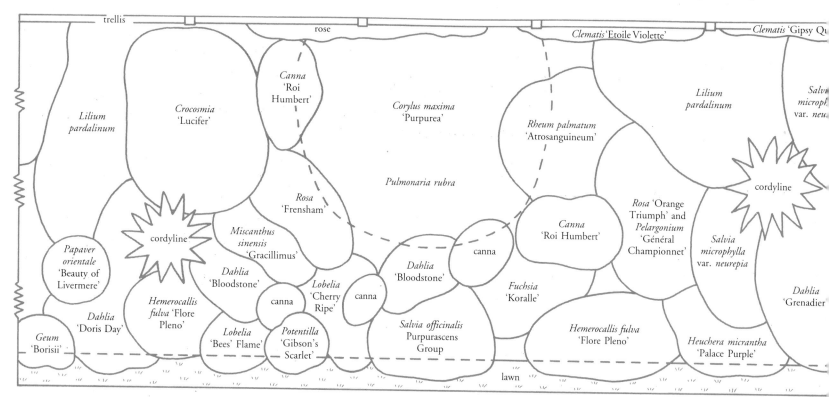

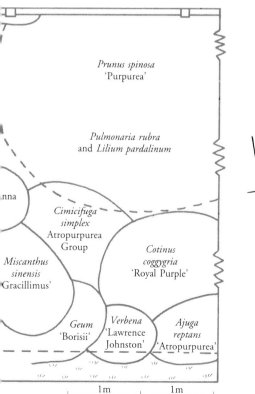

Prunus spinosa
'Purpurea'

Pulmonaria rubra
and *Lilium pardalinum*

nna

*Cimicifuga
simplex*
Atropurpurea
Group

*Miscanthus
sinensis
*'Gracillimus'

*Cotinus
coggygria*
'Royal Purple'

Geum
'Borisii'

Verbena
'Lawrence
Johnston'

*Ajuga
reptans*
'Atropurpurea'

N

1 m	1 m
3ft	3ft

ABOVE AND LEFT *An east-west rather than north-south orientation offers the best opportunities for magical lighting against the sun when it is low in the sky. Double borders can make the most of this: with careful positioning, the morning sun will illuminate first one border, then the opposite one; later in the day, looking towards the sunset, the display is repeated. Red and orange glow more intensely than other colours, though for this effect borders should not be sited where the sun is blocked out by tall buildings or trees.*

Particularly if borders are steeply banked, such orientation has the disadvantage that one side is sunny, the other shady. This can necessitate different planting for opposite sides. At Hidcote, most of the tender perennials needed for bright colour demand good light; in the shadier border shown on plan, they are concentrated in front where the light is best, heightening even more the contrast between a vivid foreground and dusky background.

The section of the shadier border shown on plan usually changes little from year to year and differs only from its earlier planting (see ABOVE *and* OVERLEAF, LEFT*) in the position of one or two dahlias and lobelias. The light foliage of* Prunus spinosa *'Purpurea' shimmers radiantly as coarser, more*

leaden purple filbert (Corylus maxima *'Purpurea', top right of the photograph) cannot. Leaves of* Miscanthus sinensis *'Gracillimus' and daylilies shine lime-green in vivid contrast to scarlet Dahlia 'Bloodstone'. Hazy flowers of* Heuchera micrantha *'Palace Purple' add a softer tone in front.* Pulmonaria rubra *gives spring colour in the shadow of the filbert and purple sloe, leopard lilies also tolerating some shade beneath their fringes. The trellis at the back supports clematis and roses, though it is scarcely tall enough to lift these above the rest of the border's planting.*

OVERLEAF *Lily-flowered tulip 'Red Shine'* (LEFT), *valued for its long and late flowering, radiates warmth against the morning sun, backed by handsome young leaves of* Rheum palmatum *'Atrosanguineum'. In early summer* (CENTRE), *Black Knight Group delphiniums add sumptuous deep tones, supplied later by* Aconitum *'Spark's Variety'; the yew hedge behind is sombre against the sun. Cordyline adds punctuation while in front a red admiral butterfly basks on a 'Frensham' rose. In midsummer* (RIGHT), *daylilies,* Dahlia *'Bloodstone',* Lobelia *'Cherry Ripe', buddleja and hips of 'Geranium' rose demonstrate the borders' colour range.*

Red', 'Olaf' and late Lily-flowered 'Red Shine' and 'Dyanito' are planted every autumn and removed after flowering to make way for tender plants.

Tender perennials are chosen to ensure rich colour until the autumn. The first to be planted out are cordylines, almost a month before other half-hardies while the tulips are still in flower. Harry Burrows selected a particularly handsome clone with broad bronze leaves and it is this that has been perpetuated at Hidcote, propagated from 'toes', fleshy suckers from which new shoots appear. Plants will last for several years overwintered under glass until they become too big. Plunged into the border still in their pots, the largest pots 38cm/15in in diameter, the work of moving them when frosts threaten is made much easier. A dummy pot is put back in each hole in the border, to save work digging new holes and to prevent tulips being planted in their place.

Johnston himself introduced tender plants to carry the display through late summer and into the autumn. Dahlias provided a major part of the show, adding bulk to the middle ranks of the border while begonias were used towards the front. One old begonia variety, 'Hatton Castle', is still used today. Graham Thomas considers Johnston selected only the choicest flowers for these borders; common annuals such as antirrhinums would have been too mundane. Dahlias are still the mainstay of the late display, though the shy-flowering varieties have been replaced. 'Bishop of Llandaff' with finely divided dark leaves is an old favourite, as are 'Grenadier' and 'Bloodstone' while the Cactus variety 'Doris Day' is a relative newcomer. The chocolate dahlia, *Cosmos atrosanguineus*, named for both its flower colour and its chocolaty scent, also finds a place towards the front.

Dahlias are lifted after the first frost or in mid autumn, whichever is soonest, the tops removed and the tubers cleaned and treated with fungicide. The roots are stored through the winter in a cool but frost-free room with no covering, and checked periodically against desiccation or rotting. The tubers are not started into growth until mid spring, potting them into 5 litre/1 gallon pots in the greenhouse. After the tulips have finished and have been removed, the dahlias are planted out, some after hardening off in a cold frame, others straight from the greenhouse.

Cannas are treated in a similar way, lifted at about the same time as the dahlias. The plants are cut down to about 30cm/12in, divided up into two or three pieces and repotted, watered once and then watered only if they look dry until the remainder of the growth withers and can be removed. In the greenhouse new growth starts to appear in late winter or early spring; the plants can then be watered more generously to provide vigorous plants for bedding out at the same time as the dahlias. Paul Nicholls now uses about fifty *Canna* 'Roi Humbert' each year, a generous quantity that makes a strong and exotic impact.

Lobelia × speciosa cultivars are hardy in this climate, though intolerant of stagnant conditions in mild winters and prone to rot if old flowering stems remain as a focus for fungal infection. The gardeners at Hidcote find it more convenient to treat them as tender perennials, lifting them at the same time as the dahlias and the cannas, splitting, potting and treating them with fungicide. The regime through the winter is the same as for the cannas, watering them only when they look dry. 'Will Scarlet', 'Bees' Flame' and 'Queen Victoria' have typically red flowers and dark foliage while 'Cherry Ripe' produces spikes of rich deep coral above soft green leaves.

Pelargoniums were among Johnston's favourite plants, though only one is found in the Red Borders, 'Général Championnet', a double vermilion Ivy-leaved variety planted to scramble into rose 'Orange Triumph'. The borders contain a considerable variety of other tender plants including sub-shrubby salvias, Triphylla fuchsias, and verbenas, among them Hidcote's own variety, 'Lawrence Johnston'. Sufficient cuttings are taken in late summer to provide stock plants, though plants for the borders are not propagated from these until midwinter. This saves space in the greenhouses and prevents the need for constantly cutting plants back. When bedding out is tackled, the plants are large enough to make a rapid impact. Because of environmental considerations, plants are raised in soil- rather than peat-based composts; Paul Nicholls finds this gives less satisfactory results, perhaps because the loam used is not good enough. Occasionally some plants of the fuchsias are kept from one year to the next to provide more bulky plants.

Some staking is needed for dahlias, delphiniums and salvias in particular. For shorter plants, sharply-forked elder twigs were traditionally used at Hidcote and worked well. The soft wood becomes harder as it ages and the few prongs can easily be worked in amongst a tangle of existing stems. However, it took some time to select and gather enough suitable twigs and they looked ugly until hidden by plants. Bamboo canes are now the only stakes used in the borders; although they have the advantage of being easy to use and not taking time to collect, they are obtrusive, a number of them remaining all too visible in late summer and autumn when dahlias and other plants are fully grown.

Garden visitors expect to be allowed to walk on lawns and think the pleasant feel of turf under foot nearly as important as its appearance as a foil for flowers. Although the climate here is almost ideal for good turf, 100,000 pairs of feet a year tread the same narrow path between the Red Borders, and the gardeners have a struggle to keep the most important section of Johnston's brilliantly designed route around the garden open at all times. Paul Nicholls blames Johnston's lack of practical experience in garden construction: having cut away the site

of the borders to level the ground, he put back a minimal depth of soil with no adequate drainage system.

Nevertheless, several measures are taken to make the turf stand up to as much wear as possible. The dwarf ryegrass used is much tougher than the traditional mix of browntop and Chewing's fescue and although rather coarser in texture, it is not unacceptably so. The lawn is cut at a height of about 37mm/ 1°in with a rotary mower, to avoid compaction, and the clippings removed. Although this height of cut seems very long for fine turf, frequent mowing will give a reasonably fine texture and the grass will take much more wear. Occasional aeration with a slitter or solid-tine machine also helps improve the surface drainage and the health of the grass. Slatted duckboards are used as a temporary measure to protect any patches of turf that are showing signs of excessive wear.

The Red Borders are a strange amalgam of superbly scaled and proportioned grand garden design with cottage garden planting, using gorgeously exotic and sophisticated flowers. Though the essence of Johnston's style of planting has been maintained, the bulk of the borders has increased and the individual plants have changed greatly as successive head gardeners and others, notably Graham Thomas, have introduced new varieties. These have lengthened the season of display, added brilliance to the foreground and provided a more effective foil of dusky foliage and flowers in the background. Few could doubt that the combination of plants is more successful today than it was in Johnston's time, a tribute to the skill of the gardeners and the benefits of continual critical reappraisal and reworking.

Two graceful pavilions backed by crisply architectural palissades à l'italienne of clipped hornbeam and billowing ilex oaks frame the termination of the garden's main axis, an elegant gateway overlooking the Vale of Evesham. Throughout the garden, Lawrence Johnston's mastery of spaces and levels is apparent. However, the scale of the planting has inevitably changed as trees such as the oaks have grown large. Thus the plants at the back have become bulkier, the rest of the planting has been edged forward, using fewer small plants, and the central grass path has been narrowed by about 1m/3ft. Nevertheless, the present position of the front of the border, in line with the outer edge of the steps, seems logical and agreeable. The gardeners try always to ensure that the outlines of the two gazebos are never hidden by the larger shrubs or small trees; skilful pruning to limit their size without spoiling their natural grace is often necessary.

BORDERS WITH ANNUALS

Le Pontrancart

Whether the required effect is Jekyllian sophistication of colour and texture or simply a glorious and exuberant *floraison*, the versatility of annuals is clearly demonstrated in the borders at Le Pontrancart near Dieppe. Careful selection of varieties, superlative cultivation and impeccable timing combine here to give widely differing borders in each of which annuals play an essential role. Their usefulness to provide colour in the summer garden, perhaps as an alternative to tender perennials, has been ignored by too many gardeners for too long.

Though a few plantsmen such as Graham Rice and Christopher Lloyd have championed annuals in recent years, there are few gardens, save for the trial grounds of the leading seed houses, that depend as heavily on annuals for their display as Le Pontrancart. Here the Old Garden consists of a labyrinth of hedged enclosures and walks lined with borders, some

one third and others entirely of annuals. The borders range from sophisticated symphonies in a restricted colour scheme to a riot of every imaginable hue; one of nothing but white cosmos edged with dusty miller shows a more simple approach, the sparkling purity of the flowers both striking and tranquil.

When the parents of the owner first came to Le Pontrancart before World War II, the Old Garden was simply the area where vegetables were grown. The hedges and flower borders appeared when the owner's father invited Kitty Lloyd-Jones to help with the laying out of the garden. Miss Jones had been responsible for improvements to a number of English gardens including that at Upton House in Warwickshire. Very much a private garden intended for the enjoyment of family and guests during six weeks or so of late summer, the Old Garden has developed as a model of French rationality, formal and

The blue and white garden at Le Pontrancart was first planted with the help of Kitty Lloyd-Jones in a style strongly reminiscent of Gertrude Jekyll. Though it relies mainly on perennials, the thirty per cent or so of annuals it contains are essential to its success. To the strong pure blue of Salvia patens *(in the foreground, on the right) are added grey-blue echinops, perovskia, nepeta and* Eryngium × tripartitum, *white anemones, achillea and cosmos and drifts of cream antirrhinums. Foliage is important too, with cream-*

*variegated dogwood (*Cornus alba *'Elegantissima') and much grey (*Artemisia ludoviciana, Stachys byzantina, Senecio cineraria *and* Tanacetum ptarmiciflorum*). Blue lyme grass, seen in the centre of the far border, is as important for its form and texture as for its glaucous colour. Rich green yew hedges and emerald turf make the perfect formal setting for generous informal planting.*

54

geometrical throughout. However, as in Claude Monet's garden at Giverny, this excess of formality has fuelled a desire for some informality to provide balance, hence a water garden laid out by the eminent English garden designer Russell Page in 1980, shortly before his death.

The owner was young when his father and Kitty Lloyd-Jones began the improvements to the garden, though he remembers the development of the blue and white garden, one of the first areas to be tackled and one that owed much to Miss Lloyd-Jones. Little changed since these early days, it remains strikingly similar to designs by Gertrude Jekyll, using few plants that she did not feature in the blue and grey plans in her *Colour Schemes for the Flower Garden*. There are even Jekyll-style drifts of some flowers such as the ivory-white antirrhinums that weave through the middle ranks of the border. The only difference is that at Le Pontrancart yellow flowers are not used.

Miss Jekyll considered some yellow – lilies, tree lupins or lemon antirrhinums – to be an essential contrast, showing off blue flowers to their best advantage, a view also held by Graham Thomas. But life, and gardens, would be very dull if we all obeyed such rules unquestioningly. Le Pontrancart's blue and white border is undoubtedly an outstanding success, gentler and more tranquil than it would be if yellow were incorporated. True, the rich pure blues would seem even more vibrant if set off by lemon flowers, but only at the expense of the soft and tranquil effect that was required.

As in Miss Jekyll's designs, the blues are mostly pure, as for *Salvia patens*, or grey-blue, as for perovskia, eryngium and echinops, only heliotropes verging towards purple and thus outside the colour range permitted by Miss Jekyll. White flowers leaven the scheme, cosmos, phlox and *Achillea* 'The Pearl' being especially striking for their dazzling purity. White crinums substitute for the Madonna

lilies used by Gertrude Jekyll for her slightly earlier-flowering schemes, with flowers of remarkable nobility and purity of line. Foliage textures are cleverly balanced, Miss Jekyll's beloved lyme grass playing a key role.

Though, as in the Jekyll designs, the annuals account for only about thirty per cent of the flowers, it is hard to imagine how the scheme could work without them. Graceful white cosmos and the repeated spires of snapdragons are essential elements of the design, while towards the front of the border the display depends largely on the likes of the blue lace flower (*Trachymene coerulea*) and *Salvia farinacea*.

Turning out of the blue and white garden, the cross axis of the Old Garden is planted with flowers of yellow, gold and lime green, all of them annuals. Here are rudbeckia, nicotiana, snapdragons, French and African marigolds, their glowing colours a telling contrast with those of the blue and white garden. Gertrude Jekyll also used this device, the blue compartment of her 'special colour garden' leading on to the gold garden and thence to a section planted with grey, lavender and pink.

The owner felt that this addition to his father's garden should be rather subdued and resisted the temptation to use a brighter and more varied mix of colours. Halfway along its length, this axis traverses the garden's main borders, planted with annuals of every hue. The possible conflict between harmonious and riotous colour schemes was resolved by isolating the two, planting, at Russell Page's suggestion, a ring of 'Iceberg' roses at the *rond point* where the axes cross.

At the end of the main borders furthest from the house is a broad exedra, an inviting bench at its centre, backed by a border of white cosmos. This is perhaps the climax of the tour of the garden, though it is much the simplest and most restrained planting here. The pristine flowers shine brilliant against the sombre yew behind, their dance in the

summer breeze echoed by the dappling shadows on the lunette lawn. It is strange that such an aristocratic flower can be so taken for granted: if it were some fractious Himalayan rarity costing a fortune for a scrap of root and impossible to grow, every gardener would want it. Because it is so cheap and obliging, white cosmos is generally ignored and seldom used to advantage. The owner's father grew white tobacco flowers here; the daily task of deadheading them, a routine for all flowers at Le Pontrancart, proved slow, sticky and disagreeable, though their scent at dusk must have been beguiling.

The main borders leading towards the house are a glorious mix of annuals in thoroughly assorted colours. Many gardeners would recognize this with nostalgia as typical of gardens of their childhood. I remember from my early days the excitement of growing exotic and entrancing annuals from small plain blue packets of seed, which never betrayed their secret potential by the lurid colour photographs of today. One never knew what to expect of some of the more uncommon species – daisies from South Africa, wildflowers from the western United States or flowers from Australia that carpet the desert after infrequent rains – described on the packet with tantalizing imprecision. Their names – monarch of the veldt,

RIGHT, TOP AND BOTTOM *A typical section of the sunnier of the two blue and white garden borders contains about four ranks of plants, including some Jekyllian drifts, in a width of 2.5m/8ft. Tall Hibiscus syriacus 'Oiseau Bleu' (to the right and left of the section on plan), echinops, aster and cosmos give height, with spikes of perovskia and snapdragon adding contrast of flower form. A proportion of solidly shaped clumps such as phlox and achillea supplies structural firmness.*

The smaller groups at the front of the border are mainly of annuals alternating with silver-leaved plants. Acanthus-like leaves of echinops, feathery fronds of cosmos, bold straps of crinum and narrower twisting, turning ones of lyme grass are also important for foliage contrast. Though not a true lily,

the crinum has lily-shaped blooms and adds an aristocratic note in the same way as the Madonna lilies used by Gertrude Jekyll. The loss of such regal flowers would substantially diminish the quality of the planting.

Seen against the light, whites and creams shine radiant while fine-textured lavender-blues such as perovskia shimmer hazily; though such colours cannot blaze with the intensity of red and orange, the effect is none the less entrancing. As at Hidcote, the ability of early morning and late afternoon sun to light the flowers from behind is enhanced by the borders' east–west orientation.

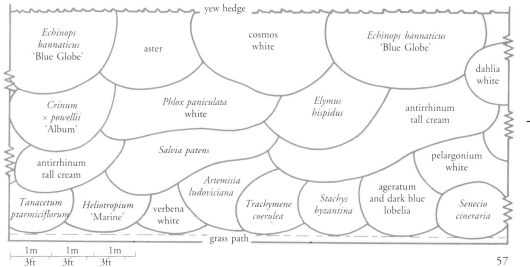

yew hedge

Echinops bannaticus 'Blue Globe'	aster	cosmos white	*Echinops bannaticus* 'Blue Globe'

dahlia white

N

Crinum × powellii 'Album'

Phlox paniculata white

Elymus hispidus

antirrhinum tall cream

antirrhinum tall cream

Salvia patens

pelargonium white

Artemisia ludoviciana

Tanacetum ptarmiciflorum

Heliotropium 'Marine'

verbena white

Trachymene coerulea

Stachys byzantina

ageratum and dark blue lobelia

Senecio cineraria

grass path

1m 3ft   1m 3ft   1m 3ft

57

California bluebell, Swan River daisy – were full of the romance of faraway places. The thrill of growing them from the tiniest seeds and of watching their first blooms unfold is vividly recaptured at Le Pontrancart when meeting annuals, such as the blue lace flower, scarcely seen on the English side of the Channel for thirty years.

Sadly, most gardeners have abandoned annuals in the endless attempt to save labour. It is perhaps fortunate that until recently France has been rather isolated from the vagaries of horticultural fashion: some of its traditions of gardening excellence, such as the use of annuals on the grand scale, have

survived here while they have been abandoned by more fickle gardeners in countries that consider themselves more horticulturally advanced. But easier plants, shrubs, perennials and ground cover, cannot provide such brilliant display in the latter half of the year.

True, tender perennials can fill this role and have once more become as popular on both sides of the Atlantic as they have always been in the favoured climates of the southern hemisphere. But there are differences: in their favour, the perennials include bigger plants of more exotic appearance; on the other hand, the annuals do not need to be propagated

from cuttings nor overwintered under glass and most are relatively cheap and easy to raise from seed. A disadvantage of annuals is that they can look disreputable as they run to seed, though this is not a problem at Le Pontrancart where the garden is used only over the month or more of late summer when they are in full and glorious bloom.

Old favourites in the main borders include zinnias, French and African marigolds, gazanias, balsam, *Phlox drummondii* and snapdragons, while at the back of the borders cosmos, lavatera, *Nicotiana sylvestris* and cleomes provide height and bulk. All the annuals and pelargoniums are raised from seed at Le Pontrancart. That they all flower together, never too soon nor too late, is the result of artifice and not luck. Nor is this successful synchronicity just a matter of staggered sowing times; much depends on the results of transplanting, an effective way of both holding them back and encouraging bushy growth.

The owner and his head gardener have come to know precisely how long each variety will take to come into flower after transplanting. Thus, choosing when to transplant from the nursery is just a matter of waiting until the right time. Some plants seem to grow beyond their ideal transplanting size but even then they should not be moved before the appointed week if they are to flower exactly on cue. There are a few plants that resent being moved and have to be sown in situ. Though transplanting is an extra job, with experience and precise timing it becomes a straightforward task and well worth the effort.

The quality of the plants produced in the nursery owes much to farmyard manure, used on a three-year cycle. In its first season, while still fresh, it provides heat to warm the cold frames during spring. In its second season, it is used to provide a rich growing medium for the young plants, which must be vigorous if they are to perform well. If bushy, fewer plants are needed, the cost of seed is cut and a longer display is ensured. Any flower spikes will be longer and so will remain in bloom for valuable extra weeks. In its third season, the manure is sieved to make compost for seed and the smallest seedlings.

The perennials that are used around the garden must all give a good display in late summer. This is generally a matter of choosing late-flowering varieties, though in some cases, as for delphiniums, the trick of cutting down early to get a second crop of flowers is used. Trial grounds and catalogues are studied each year for new annuals that

might be used, though the changes tend to be few. Catalogues are treated with caution, particularly those showing flowers of a glorious blue that turn out to be an indifferent mauve. One year's trial is essential to be sure that any newcomer is truly worthy of a place in the borders. At the onset of autumn, the owner and his head gardener review the year's successes and failures in preparation for the next seed orders, which they obtain from leading seed houses in France, England and the United States.

The yew hedges which provide the framework for the garden were originally low so that the entire Old Garden could be seen at once. Though this was undoubtedly spectacular, there was not the least element of surprise. The owner thinks that short hedges were not Miss Lloyd-Jones's ultimate intention: she merely wanted them to grow strong and bushy. During World War II, under German occupation, the borders

LEFT *The Old Garden is bounded by a wall to the east and, hidden by hedges towards the top of the picture, a canal to the south. The blue and white garden (on the right) lies in the corner nearest the house. There are borders of dahlias edged with annuals parallel to it, separating it from the garden's main borders. The yew hedges that frame the garden are square topped and of even height, save for arches into each of the garden's most intimate enclosures, with ball finials at the rond point. A small garden bordered with coloured-leaved shrubs is beyond the main axis to the right of the picture.*

RIGHT *As the sun dips behind the encircling hedge, at the far end of the main border, the contrast between shining white cosmos and black-green yew is at its most dramatic. Grey-leaved dusty miller (Senecio cineraria) covers the bare ankles of the cosmos while the verdant lawn provides an arena for the play of shadows. Where sparkling purity is at a premium, there is no need for the seat to be anything other than the most brilliant white; a broken tone would diminish the effect. Daily deadheading ensures that the display is never marred by faded blooms and encourages continual flowering.*

contained only potatoes and the hedges were neglected. When flower gardening started once more, the hedges had become big enough to separate each area of the garden. After the war, the owner's father's enthusiasm for gardening had diminished but Miss Lloyd-Jones was invited back to help improve the blue and white garden. Spurred on by this one great success and by the lack of anything but lawn within the rest of the hedges, his enthusiasm returned. One by one the borders were replanted, something the owner thinks might never have happened had not the hedges remained to draw attention to the lack of flowers.

The owner claims little credit for the development of the Old Garden. This modesty is unjustified, for he has clearly refined and improved every part of it. However, one significant change due to neither his father nor Miss Lloyd-Jones is a striking pink and red border, running within the garden wall nearest the house. Originally this contained mixed colours which the owner thought rather dull. He wanted to include all shades from orange-red to purple-red, a combination many would consider strident and unappealing. He too was apprehensive but soon found that by including enough green foliage, what could

be the most glaring clashes could be transmuted into vibrant harmonies. The inclusion of pink and the very sparing use of maroon foliage prevent the colours from seeming the least bit leaden.

Perhaps two thirds of the plants in the pink and red border are annuals. To achieve a full border with heights rising to 1.5–1.8m/ 5–6ft, single-colour selections of some of the annuals are needed. These must include taller varieties of, for instance, antirrhinums, no longer so easy to find as they were thirty years ago. Seed houses today often seem to favour shorter varieties suitable only for carpet bedding and mixtures rather than single colours, in spite of the constant clamour from gardeners for varieties they can use in fuller borders such as this with a distinct colour theme. Touches of salmon, particularly from tall antirrhinums, relate the border to the soft tones of old brick of the garden walls and house.

The pink and red border is perhaps the most original of all the borders at Le Pontrancart, a daring combination of vivid tones found in few other gardens. Almost every border here depends on annuals, used in a variety of ways. Whether it is the simplicity of little more than white cosmos, the refinement of a Jekyllian scheme interpreted by Miss Lloyd-Jones, the panache of pink and red or simply a carefree riot of charming and old-fashioned flowers, each border is a compelling demonstration of their usefulness. The time has come to re-examine the role of annuals in our gardens and learn from the abundant successes of Le Pontrancart.

LEFT *The red and pink border is perhaps the most innovative of those at Le Pontrancart in that it breaks the rule that red containing blue should not be mixed with red containing orange. Such a combination could seem uncomfortably discordant but potentially violent colour clashes are tempered by plentiful green foliage to give a dazzling and vibrant display. Cerise and pink phlox jostle with scarlet* Salvia coccinea *'Lady in Red' and tall mauve-pink* Lavatera *'Loveliness' rubs shoulders with fiery red and rose dahlias to either side. Occasional touches of dark foliage add weight and punctuation while the whole is set off by a shimmering background of white poplar.*

RIGHT *A view along the main borders towards the semicircular lawn and the white bench shows their thoroughly mixed flower colours. Many of the annuals are favourites of today, such as French and African marigolds, nicotiana and snapdragons, but there are a few popular plants of the past, such as coreopsis and balsam* (Impatiens balsamifera). *'Iceberg' roses surround the* rond point *where the cross axis intercepts.*

A PURPLE BORDER

Sissinghurst Castle

The length of display in almost all parts of the garden makes Sissinghurst in Kent unique. The precise use of colours, the plant associations devised and the frequency with which they are recombined serve to make it a stimulating garden to visit time and again. Few gardens have acquired such a potent mythology as Sissinghurst, the inevitable result of a steady stream of television series and books about the garden's creators, Vita Sackville-West and her husband Sir Harold Nicolson.

Many now see the garden as an immutable and inviolable shrine to their memory. But as John Sales, Chief Gardens Adviser to the National Trust, has pointed out, a garden is not an object: it is a process and cannot be frozen at a single moment. This process, the evolution of Sissinghurst, has continued without cease, enriching the garden within the framework set down by the Nicolsons by the addition of more and longer-flowering plants and by the application of the most artful horticultural technique.

The Purple Border was probably the first to be made in this colour range. It is curious that purple should be so little used. In the home, rooms of red, yellow, green or blue are not uncommon but purple is seldom seen. Outside, too, it is undeniably a difficult colour. At Sissinghurst, the aspect of the border is such that purples look sullen and sulky in the middle hours of sunny days, just the time when the garden is open, though the soft light of early morning or dusk, or cloudy weather, is more flattering. Gertrude Jekyll disliked the colour and especially hated magenta. Graham Thomas thinks this may be exactly why Vita chose it. He has observed that in Miss Jekyll's day, few of the silver-leaved plants, such as artemisias and helichrysums, that we use to soften or leaven

The border in midsummer, before the hips of Rosa *'Geranium' turn scarlet. Its success depends on lighter and brighter tones as well as the deepest, richest purple. Airy lilac* Thalictrum delavayi, *silver* Eryngium × tripartitum, *becoming blue as the flowers age, and* Monarda *'Beauty of Cobham', its pink flowers set off by purple calyces, provide the paler tones while magenta* Lythrum salicaria *'Robert' and* Liatris spicata *add bright colour. The richness of* Dahlia *'Edinburgh' is enlivened by its white-tipped* petals. Allium sphaerocephalon *and* Lavandula *'Hidcote' add red- and blue-purple while bass notes are provided by a background of smoke bush* (Cotinus coggygria).

purple and magenta flowers were known or available; it is only since World War II and the pioneering work of gardeners such as Mrs Desmond Underwood and Graham Thomas himself that silver has come to be widely used.

At Sissinghurst, the problems of purple have not been resolved only by using silver: the process by which leaden violet and screaming magenta have been made to yield a border of richness seasoned with vibrancy, softened by silvery lavender blue and lilac pink, and the techniques used to wring every last scrap of horticultural effect from a difficult site are object lessons to every gardener.

Great works of art often succeed because of the tension between opposites, Apollonian order and control versus Dionysian exuberance and excess. This is often so in gardens, many of us preferring an ordered framework holding informal and overflowing planting. Yet most gardeners would prefer the framework to be less Apollonian than Versailles and the planting to stop short of Dionysian riot. At Sissinghurst, Sir Harold restored order to a chaotic site, uniting the remaining fragments of the sixteenth-century castle into a simple and well-scaled formal design while Vita supplied the planting.

Those who remember the garden in her day agree that she had a marvellous vision of the abundant effect she wanted to achieve but not all feel her planting succeeded: to many, swayed by her writings, it was gloriously romantic. But to most professional gardeners it was Dionysian to a fault, lacking the horticultural control needed to maximize its beauty, lengthen its season of display and ensure a balance between key plants.

One suspects a slight element of sour grapes in the professionals' reservations about the garden. But then the Nicolsons were unashamedly amateur: they had no need to make a garden that pandered to the public's demands for continual display in every area from spring to fall. They had the Lime Walk for spring, the White Garden and Rose

Garden for early summer, asters and dahlias for late summer and early autumn – more than enough beauty to satisfy a single family at every season throughout the gardening year. If one area had passed its peak, there was always another part of the garden to enjoy. It did not matter that standards of construction and maintenance might be a little less than perfect here and there, nor that there were a few weeds. Sissinghurst was already a garden of ample delights.

Pamela Schwerdt and Sibylle Kreutzberger were employed by the Nicolsons in 1959, two and a half years before Vita's death. On their first visit on July 17, Vita's comment that 'when I see the [*Alstroemeria*] Ligtu hybrids I know the season is over' was taken

as a challenge to make the garden as full of flower in the latter half of the year as it was earlier. The Purple Border then was not at all as we see it today: *Rosa moyesii* and its hybrid 'Geranium' survived from an earlier scheme, with *Thalictrum delavayi, Allium stipitatum, A. sphaerocephalon*, white-tipped *Dahlia* 'Edinburgh', purple *Cotinus coggygria* and Michaelmas daisies, some of them inferior seedlings, supplying the purple and mauve.

One iris and several violettas at the front of the border started the display but there were no earlier flowering plants. Miscellaneous seedlings of *Campanula lactiflora* lightened this sultry scheme, with electric blue *Anchusa* 'Loddon Royalist', close in the spectrum to purple yet far from

harmonious, adding a tone that might equally be considered enlivening or disagreeable. On the wall, vines were more in evidence but the range of clematis was limited to about three, including silvery *C. × jouiniana* 'Praecox'. Perhaps only about one-third of the plants grown there now were used in the Nicolsons' time and the intensity and length of season of the display were far less than we see today. The garden has been enriched in recent years by many plants that were simply not available when it was first made.

Vita tended to buy plants in ones (Sir Harold was more prodigal and bought in threes or fours) and was pleased if they spread themselves generously by seed or root, considering them to be good value. Such

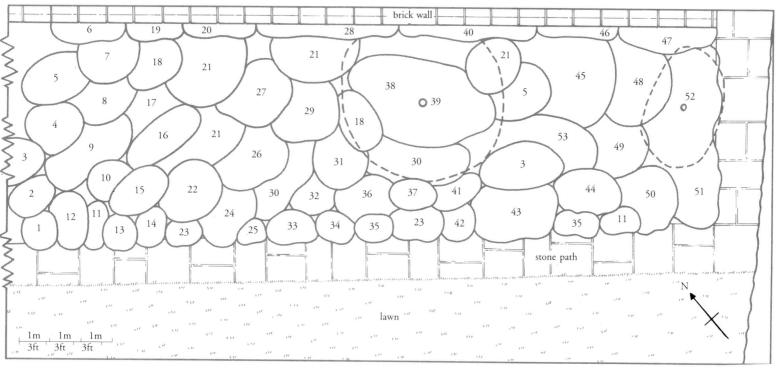

brick wall

6   19   20   28   40   46   47

7   18   21   21

5   18   21   21   45   48

8   17   27   5   52

29   38   39

4   16   21   18   53   49

3   9   26   30   3

10   31   30

2   15   22   44   50   51

12   11   24   30   32   36   37   41   43   35   11

1   13   14   23   25   33   34   35   23   42   35

stone path

N

lawn

1m 1m 1m
3ft 3ft 3ft

LEFT *The border in midsummer. Lilac-pink*
Alyogyne hakeifolia, *a recent addition, its buds still
furled in early morning, appears in front of*
Clematis × durandii. *Though lax and usually
trained on a wall, the clematis is here grown as an
herbaceous plant supported on stakes. Statuesque
cardoons at the back match the height of the roses.*

KEY TO PLAN
1 *Dianthus* 'Annabel'
2 *Dierama pulcherrimum*
3 *Eryngium × tripartitum*
4 lupin 'Blue Jacket'
5 *Delphinium* Black Knight Group
6 *Clematis* 'Niobe'
7 *Cynara cardunculus*
8 *Dahlia* 'Good Intent'
9 *Salvia × superba*
10 *Aster novi-belgii* 'Carnival'
11 *Salvia lycioides*
12 *Iris* 'Sissinghurst'
13 *Sedum* 'Vera Jameson'
14 *Erysimum* 'Constant Cheer'
15 *Monarda* 'Vintage Wine'
16 *Phlox paniculata* 'Cool of the Evening'
17 *Aster novi-belgii* 'Climax'
18 sweet pea 'Noel Sutton'
19 *Clematis* 'Abundance'
20 *Vitis vinifera* 'Purpurea'
21 *Malva sylvestris* var. *mauritiana*
22 *Geranium psilostemon*
23 *Verbena* 'Silver Anne'
24 *Aster × frikartii* 'Wunder von Stäfa'
25 *Petunia* 'Purple Defiance'
26 *Baptisia australis*
27 *Delphinium* Blackmore and Langdon's Hybrids
28 *Clematis* 'Perle d'Azur'
29 *Dahlia* 'Requiem'
30 *Dahlia* 'Pink Michigan'
31 *Aster novi-belgii* 'Cliff Lewis'
32 *Cleome hassleriana* 'Violet Queen'
33 *Salvia viridis* 'Bluebeard'
34 iris
35 *Dianthus amurensis*
36 *Liatris spicata*
37 *Platycodon grandiflorus*
38 *Stachys macrantha*
39 *Rosa* 'Geranium'
40 *Clematis* 'Etoile Violette'
41 *Allium sphaerocephalon*
42 *Lavandula angustifolia* 'Hidcote'
43 *Geranium × magnificum*
44 *Iris sibirica* 'Keno Gami'
45 *Thalictrum delavayi*
46 *Clematis* 'Ernest Markham'
47 *Clematis × jouiniana* 'Praecox'
48 *Allium stipitatum*
49 *Lythrum salicaria* 'Robert'
50 *Monarda* 'Beauty of Cobham'
51 *Hosta ventricosa*
52 *Cotinus coggygria* 'Foliis Purpureis'
53 *Dahlia* 'Edinburgh'

OPPOSITE *Much of the border's charm derives from its
varied group sizes:* Rosa *'Geranium' and* Stachys
macrantha *towards the back are about 5m/17ft
across, while groups of tiny plants in front are as little
as 60cm/2ft wide. The wall is planted to be almost
entirely covered in summer, clematis accounting for
more than three-quarters of the varieties. The full
planting masks the border's tapering shape.*

ABOVE *Vibrant magenta* Liatris spicata, Monarda
'Prärienacht' and Lythrum salicaria *'Feuerkerze'
dominate the duskier tones of* Clematis *'Perle d'Azur'
and C. 'Etoile Violette'. Silvery* Campanula lactiflora
*is an essential element of the scheme, preventing it
from appearing leaden by adding a lighter note.*

plants added to the overflowing abundance of the garden; but there were many inferior offspring and some proved to be thugs, ousting choicer and less competitive plants whose many labels remained as epitaphs. Pam and Sibylle's earliest task was to rogue out unsatisfactory seedlings, reduce overlarge groups and propagate those plants that had become too small and insignificant to play their required role.

They encouraged Vita to widen the range of plants, adding *Campanula glomerata* 'Superba', *Eryngium × tripartitum*, choicer Michaelmas daisies and *Salvia × superba*; *Persicaria bistorta* 'Superba' (syn. *Polygonum bistorta* 'Superbum') was a daring innovation, its silvery pink with a hint of lilac a break from the existing colour scheme. Vita herself brought back four clematis from Christopher Lloyd's nursery at Great Dixter nearby, an extravagance she thought very dashing. Three of these, 'Perle d'Azur', 'Etoile Violette' and 'Ville de Lyon' survive on the wall, though 'Star of India' needed a different pruning regime and is no longer grown here. Anchusas were soon phased out, to be replaced by flowers of more sympathetic hues, not only rich purples but vibrant magentas such as *Geranium psilostemon* that dominate from early to mid summer. The Moyesii roses, valued for the contrast of their dusky red flowers and scarlet hips saving the border from tameness, were pruned to reduce their sprawl, allowing the planting to continue beneath their gracefully arching stems.

Other important additions to the scheme were sword-leaved *Gladiolus byzantinus* and the spikes of delphinium for contrast, *Stachys macrantha*, platycodons, *Lobelia × gerardii* 'Vedrariensis' and phlox, plus some good irises moved from the Rose Garden, all newcomers being placed so that those flowers that were out at any one time were not bunched together in a single part of the border nor so spread out that the interplay of colours did not read from one group in

flower to the next. A wide variety of shapes and textures is felt here to be crucial to the success of a border if such a limited colour range is being used: 'mounds and fluffs' such as *Thalictrum delavayi* and Michaelmas daisies are used to separate plants with more definite form such as delphiniums or cardoons, allowing each of these handsome brutes the space to express its individuality.

Following Sir Harold's death in 1967, Sissinghurst was donated to the National Trust. The gardeners were relieved that the Trust made no attempt to ossify the garden by decreeing that nothing should change. As Pam Schwerdt says, 'had Lady Nicolson been alive, she would always have been adding plants. We were so thankful that somewhere along the line somebody decided that Sissinghurst was going to be a place where we would go on adding rather than that someone would absolutely stop the clock.' Nevertheless, nobody wanted to change the character of the garden. This policy of flexibility within unchanging guidelines has proved a creative stimulus, allowing the garden to be constantly refreshed and renewed from year to year.

The border used to be 1.2-1.5m/4-5ft narrower at the top end. In the early 1970s it was widened, making possible much more generous planting within. This had the additional benefit of allowing the border's stone edging path to align more satisfactorily on the library door. At the same time, the level of the lawn was raised so that it was flush with the path, allowing mowing over the edge and obviating the need for an unsightly and labour-intensive gulley, one of many simple changes to the structure of the garden which have both improved its appearance and reduced work.

Pam Schwerdt and Sibylle Kreutzberger regard the Purple Border as being much easier to plant than the White Garden: whites tend to be either pure, grubby or yellowy and no two such sorts will flatter one another,

whereas far more shades of purple will mix agreeably. Some people, including Christopher Lloyd, feel that the border is too subfusc, though much has been done to try to prevent this, adding just enough lavender blues (for instance *Clematis* 'Perle d'Azur') and pinks (such as *Persicaria bistorta* 'Superba') to leaven the scheme without destroying its richness, and magenta (for example, liatris and *Geranium psilostemon*) to give sparkle.

One of the first tasks of the year is the pruning of the clematis, at Sissinghurst tackled at any time between November and February: as almost all are Jackmanii or Viticella varieties, this is a fairly standard procedure with no need to worry about different pruning strategies, all of them being cut to about 30cm/12in from the ground. The wall is covered with 15cm/6in mesh galvanized pig wire wrapped over its top so that the uppermost growth remains firmly attached and does not form an unruly quiff. The vertical strands of the pig wire give it an advantage over more traditional horizontal wires that allow the clematis to slip from side to side and do not give it the firm anchorage it likes. In spite of its inelegant name, the pig wire is perfectly neat and unobtrusive throughout the spring while the clematis makes its annual journey up the wall.

When covered in late summer bloom, the wall is an essential element of the border, allowing the display to continue upwards. In terraced gardens, as at Powis Castle, walls allow the colour scheme to carry even further up, continuing from border plants via wall planting to border plants on the level above.

The secret of Sissinghurst's success with clematis lies in constant training and tying as the plants regrow through spring and early summer: every couple of weeks, the new shoots are spread out and tied in, using wire twist ties. By midsummer they cover almost the entire wall. This strict regime prevents the clematis becoming a vertical lump of tangled

stems but it is also beneficial in discouraging clematis wilt. This omnipresent and opportunistic disease lurks in wait ready to enter any stem through a wound caused by bending or kinking and can reduce plants to a shrivelled mass in hours; clematis that are firmly attached cannot bend or kink and seem relatively free from attack. The advice from Sissinghurst is never to let your clematis swing about in the wind, nor to let them get so tangled that they need to be unravelled and to handle them as little as possible.

In almost any border planned for summer display, there is ample scope for spring flowers too. The Purple Border makes use of wallflowers, not just the perennial sorts such as *Erysimum* 'Bowles' Mauve' and *E.* 'Mrs L. K. Elmhirst' but those raised as biennials, particularly *E.* (formerly *Cheiranthus*) 'Ruby Gem' (syn. 'Purple Queen'). Sissinghurst's

present Head Gardener, Sarah Cook, considers the traditional method of raising them by sowing in nursery rows and transplanting to be a waste of both time and seed, though the inevitable damage to the root system on transplanting does have the effect of encouraging bushy growth. At Sissinghurst, seed is sown in pans in mid June, seedlings being pricked out into plugs before being transplanted into the nursery in late July. Though the sowing is late, plants do not suffer a check on being planted in the nursery. When ready to move to the border in November, the plants are a little smaller than usual but Sarah considers this an advantage as the traditional method can give plants that are too big and, if they are not generously spaced, too leggy. Sweet rocket is also used through the border, grown in the same way and planted particularly towards

ABOVE, RIGHT *In spring, tulips occupy gaps in the border that will later be filled by adjoining plants or by tender perennials. Here optimistically named 'Blue Parrot' (it is mauve) and 'Dairy Maid' are complemented by more richly coloured dwarf iris 'Sissinghurst' and sweet rocket (Hesperis matronalis).*

ABOVE, LEFT *Dwarf iris 'Sissinghurst', wallflowers and sweet rocket continue the display after the tulips fade. Labels mark only those plants in or close to their flowering season. The wall, yet to be covered as clematis grows, is unobtrusively wired over its top with pig netting.*

OVERLEAF *In late summer, the wall is a mass of interwoven colour. Small and silvery flowers of* Clematis × jouiniana *'Praecox' leaven the deeper colours of* CC. *'Victoria', 'Leonidas' and 'Mme Julia Correvon'. Though the sparse and stiffly borne leaves of* Malva sylvestris var. mauritiana *may seem ungainly, its repeated vertical accents and rich magenta flowers are an undoubted asset.*

69

the back. When the sweet rocket and wallflowers approach the end of their display in late spring, they are replaced with tender plants including dahlias.

Tulips are well represented: 'Blue Parrot', white-edged light purple 'Dairy Maid' and amethyst 'Attila' are old favourites but perhaps the most striking is 'Greuze' in particularly glowing violet. Plants such as these that have a short season are labelled only while in flower: when the petals drop, the labels are whisked away, minimizing the baleful effect of a host of black plastic squares covered with white letters, a regrettable necessity in a garden filled with so many fascinating and unusual plants.

There can be few gardens where staking is practised so unobtrusively and effectively. The structure of the garden makes staking essential: turbulent gusts set up by the courtyard walls and the high tower are enough to flatten many plants that are normally upstanding. In the Purple Border, plants that in most gardens flop or snap or have to suffer the indignity of being uncomfortably trussed are supported using a variety of techniques, so invisibly that few would realize the massive quantity of brushwood within. Hazel is much the most popular material, being far less noticeable than birch. It is grown in an out-of-the-way corner of the garden and coppiced on a four-

By early summer the forest of brushwood, essential support for sprawlers such as Clematis × durandii *and unnaturally top-heavy skyscrapers is virtually hidden. Here* Delphinium *Black Knight Group is able to sway gently within hazel stakes without suffering the indignity of being trussed tightly to bamboo canes.* Rosa *'Geranium' still bears the last of its flowers and deep lavender blue* Galega orientalis *peeps out from behind the clematis.* Erysimum, Nepeta sibirica, *ruby-purple rose 'News' and the deep crimson pincushions of* Knautia macedonica *fill the furthest end of the border while* Geranium psilostemon *dominates the foreground.*

year cycle, giving brush of just the right thickness and length.

Stakes are inserted when the plant has reached about two-thirds of its final size, staking 8-10cm/3-4in over the plant's head. The tops of the brushwood are generally bent over to make an interlocking mesh that disappears in as little as a week as the plants grow. However, such rigid staking does not suit all plants, particularly skyscrapers such as delphiniums and *Thalictrum delavayi* that need to sway in the wind and will snap at the point of support if too firmly held; for these, the tops of the brushwood are not bent over. The stems can still move about, helping them to shed some of the weight of rainwater that is so often the cause of their collapse.

It is obvious and essential that stakes should be pushed well into the ground if they are to give adequate support, though failure to do so is all too often the cause of ignominious collapse. If the ground is too hard, it should be watered first. Great care is taken to stake each clump at the exact moment when it has reached the right size, and only three or four groups are tackled at a time. There is no question of staking one-third of the border plants all at once and having to see a mass of brushwood above some of the clumps for longer than is absolutely necessary. To start with, two or three stakes are placed in the middle of the clump to keep the central stems in position. Then further brushwood is worked into the perimeter of the group. The brushwood can be used to adjust the size or the position of the clump, easing out the stems to make it seem larger or moving the stems to one side if they are not quite in the right position, effectively edging over the whole group.

Plants such as onopordum tend to topple at ground level; for these, the stem is firmly attached low down to a stout stump hammered into the ground. Something like *Salvia × superba* is supported with those thick ends of hazel that remain after the feathery

parts have been cut off for use elsewhere. A few such sturdy fans will prop up a substantial group but for plants like these with wiry stems, there is no need to stake to the top of the clump. Furthermore, especially for plants like the salvia that will continue to flower well if deadheaded, it is important not to stake so high into the clump that deadheading reveals the stakes.

Plants like *Geranium × magnificum* and aquilegias, that do need some support to the top but have to be deadheaded, pose a problem: if the flowering part of the stems is removed, the stakes are revealed. The solution at Sissinghurst is to cut the whole clump down to the ground after flowering. New leaves are rapidly produced and the clump is restored to respectability for the rest of the summer. Plants like *Geranium psilostemon* and *Galega orientalis* are cut back before they have quite finished flowering. Groups are cut only one or two at a time, so that there are never too many bare spaces; those gaps there are should never be next to each other.

Regular replanting of groups is essential to maintain balanced scale as well as ensuring vigorous plants, longer flower stems and thus extended flowering. At Sissinghurst, wholesale replanting of a large proportion of the border is felt to be too disruptive: new plantings would not achieve their full potential in their first season and the replanted part of the border would look thin. The solution is to replant just individual groups. Asters are replanted every other year but some plants are left much longer. For plants like daylilies, hostas, *Geranium psilostemon* and *G. × magnificum* that take a full season to reach a good flowering size, a large piece is carved off immediately after flowering, fattened in a pot for eleven months and returned to replace the largest and most congested of its type in the border in time for its yearly flowering. Such unseasonal planting is feasible for pot-grown plants provided they are transplanted in moist weather or watered

in well after planting.

Irises are an important element of the display of most parts of the garden, including a considerable number of Intermediate Bearded cultivars. These have the advantage of being short enough to avoid being clipped by visitors' bags yet are still tolerably graceful, something that cannot be said of all the Dwarf Bearded varieties. Though they all appreciate thorough replanting from time to time, a dodge has been devised at Sissinghurst to avoid such disruption and save a good deal of work. After flowering, the sections of the rhizomes that have flowered are chopped away with a trowel and the soil improved with organic matter for the vigorous young sideshoots to root and grow into. For those varieties that do need to be replanted, pot-grown offsets are reared in the nursery until they are big enough to be planted out.

The orientation of the iris fans is important: they should not all aim one way, towards the path or under a neighbouring group of plants, but should point in varied directions so that even after they have made some growth, their allotted space is evenly filled with shoots. Particularly near to heavily used paths, taller flower stems are often knocked over by visitors and have to be staked with 60cm/2ft split bamboo canes.

Every year a handful of the border's permanent plants is moved. Sarah Cook finds this one of the hardest tasks: putting a single group in a new position can throw out a whole series of colour relationships; these must be rectified without rearranging half the border. Perennials are moved into spaces where half-hardy plants have grown for a few years, where the soil has been much improved by the addition of compost.

A few new plants are introduced each year; in 1993, a tall mallow, *Malva sylvestris* var. *mauritiana*, with black-veined magenta flowers and lilac-pink *Alyogyne hakeifolia* were added. Both have a certain rugged charm; one suspects that Pam Schwerdt and Sibylle

ABOVE *Few flowers can match clematis for furnishing abundant colour with poise and elegance. Every bud nods gracefully, every bloom seems perfectly positioned, its slightly downturned face looking the viewer squarely in the eye from above. Here 'Perle d'Azur' and 'Etoile Violette' supply a colourful background for lively magenta* Malva sylvestris var. mauritiana.

OPPOSITE *In early autumn,* Rosa *'Geranium' casts a veil of shimmering scarlet behind, while froths and fuzzes of asters give mists of mauve and purple in front: here are* AA. × frikartii *'Wunder von Stäfa',* novi-belgii *'Cliff Lewis' and* turbinellus. Lavandula *'Hidcote' has been cut down the instant its flowers have faded, rather than in spring; the foliage is quickly renewed and the bush prevented from becoming leggy.*

Kreutzberger would consider them not sufficiently *soigné*, though Vita might well have approved. Another newcomer in 1993 was *Salvia lycioides*, its dusky purple flowers charming without being showy; Sarah Cook likes to include a few little-known rarities and the visitors certainly seem to enjoy discovering them.

Rarities and commoners alike need to be pest-free. Because roses are important in several parts of the garden, including the Purple Border, a modicum of spraying is essential. Sarah Cook considers it is not acceptable to have the plants defoliated by blackspot or rust, nor should they be blighted by aphids. The roses are sprayed about once every three weeks with a fungicide that is also effective against mildew on asters and monardas. If a fungicide such as triforine is being used, the mallows are sprayed against rust, too. Other pests in the Purple Border that are regularly controlled are blackfly on the cardoons and capsids on caryopteris, while tarsonemid mites, which disfigure the flowers of asters, are tackled by a monthly spray from late spring onwards. Slugs are a problem after a succession of mild winters but an occasional application of slug pellets or a spray with a formulation of aluminium sulfate, copper sulfate and sodium tetraborate usually restricts them to an acceptable level.

Each year when the border is being worked through, generally in early to mid spring, an ample dressing of well-rotted compost is worked in, though it is simply applied as a mulch over the roots of clematis at the back. Either a general fertilizer or one with added trace elements is applied each spring. However, Sarah Cook is reluctant to overdo the trace elements: in recent years, leaf distortions and discolorations on numerous plants at Sissinghurst have been shown to be due to manganese toxicity, salts of this metal being released from the soil as a result of years of generous fertilizer application.

This is not a labour-saving garden: Sarah

Cook is scathing about visitors who long for their own Sissinghurst and garden designers who inflict their version of it on unwitting clients. 'They have no idea what they're letting themselves in for,' she says. 'The only way to achieve a garden like this is by constant hard work, endless observation and criticism. If you plant in this way but are not willing to make that sort of commitment, you won't get a garden like Sissinghurst. It's better, and less disappointing, not to try.' Yet this warning will not deter any of us from enjoying Sissinghurst, trying to learn from it and copying at least a few of its plant associations, practical techniques or colour schemes.

In the Nicolsons' day, the garden at Sissinghurst had good bones, effusive – though perhaps not always exactly chosen or widely varied – planting and abundant romance. Although Vita Sackville-West's articles had captivated the public's imagination, it was scarcely then a great garden: it had little originality and a relatively short season of display; its standards of maintenance and construction were less than perfect. There were plenty of other gardens which, though not so much in the public eye, were equally romantic. In the thirty-odd years since Vita's death, it has been refined and enriched by a team of gardeners, led for most of those years by Pam Schwerdt and Sibylle Kreutzberger. The continuity provided by the skill and artistry of this remarkable team has allowed the garden to mature and fulfil, and even to exceed the vision of the Nicolsons. The evolution of the Purple Border demonstrates how the subtle refinement of a difficult colour range, the lengthening of the season of display and the development and application of skilled horticultural technique can be achieved. We should not forget the constant reworking over many years, the self-criticism and hard labour involved, and be grateful to all its creators that Sissinghurst survives triumphantly for gardeners to enjoy.

CONTRASTING DOUBLE BORDERS

Cliveden

The twin herbaceous borders at Cliveden in Buckinghamshire are fraternal, not identical: in summer one side burns with reds and yellows while the other soothes with cooler pinks and blues. Though their scale is large, the principles behind the design apply to more modest domestic gardens, not least because the vast area of Cliveden makes such demands on its National Trust gardeners that the borders must be simple to maintain.

These borders, facing each other across Cliveden's grand entrance court, are considered by Graham Stuart Thomas to be as good as any he has designed and the most true to his original plan and intentions. Coming from perhaps the most skilled and influential designer of planting since Gertrude Jekyll, this makes them not only of considerable historic importance but of immense value as a demonstration of the way ideas about planting have changed since Miss Jekyll's day.

The borders were planned in 1969 and planted over the next few years. Except in the spring, when both borders contain yellow daylilies and blue geraniums, they follow Graham Thomas's colour theory of 'the omission rather than the inclusion'; he explains, 'One border has got no pinks, mauves or purples, the other has got no oranges, reds or yellows; and that, I think, is the way to do gardens.'

The scale of the entrance court at Cliveden is such that very large plant groups are needed: most in the middle or towards the back of the border are 1.8–2.4m/6–8ft long,

The generous width of the soft-coloured border demands a backing of the tallest and most imposing plants such as plume poppies and onopordums. Along the front in late summer, mounds of sedum and 'Jackman's Blue' rue are contrasted with spikes of Salvia nemorosa 'Ostfriesland', grassy Hemerocallis lilioasphodelus (syn. H. flava) and the bold foliage of bergenias. The borders' designer Graham Thomas considers that white and the palest pink are needed in quantity on this side to give it impact when seen from a distance and to match its weight against that of the hot-coloured border opposite. Shasta daisies were meant to perform this function but have not thrived and only a few remain; their role is performed instead by Achillea ptarmica 'The Pearl'.

though a few are as long as 3.6m/12ft. The majority of plants used are coarse-textured and bold so that they appear to be neither amorphous seen from a distance nor interminably tedious at close range. From the drive some 70m/75yds away, groups of this size are big enough to make an impact and perfectly in scale.

Graham Thomas acknowledges that Gertrude Jekyll has been the main inspiration behind his designs for border planting and considers her *Colour Schemes for the Flower Garden* to have been his principal source of knowledge. When he visited Miss Jekyll at her home, Munstead Wood, in September 1931, he thought the main border there to be 'better than any border that I've seen since, bolstered with big bushes at the back with great clumps of yuccas and bergenias; it was altogether richer, more filled, more bulky.' On that occasion, Miss Jekyll showed herself to be little interested in botanical rarities,

preferring plants that made a definite effect. Most of her designs rely on a fairly limited repertoire of well-tried favourites; although such a restricted range of plants might seem dull to the plantsman, this gives her schemes much of their individuality, each plant being just about the best sort to achieve a particular effect. The main border at Munstead was planned to provide display from mid to late summer, something Graham Thomas considers much easier than the spring to mid autumn borders that most garden visitors expect.

In spite of his debt to Miss Jekyll, Graham Thomas's style of planting differs in many respects. While Jekyll often produced colour schemes based on a spectrum or a single colour, Graham Thomas prefers a less 'forced' system, his 'colour scheming by omission'. Thus he might choose to leave out strong yellows, or perhaps hot colours, but examples in which he has used a spectrum or a single colour are rare. He has made wide use of

silver foliage, not just to harmonize with pastel colours but to leaven richer tones. Renowned as a plantsman, he always chooses the best plant for each association in exactly the right hue and allows himself a far wider palette of plants than did Miss Jekyll. This approach has allowed him to emphasize the individuality of each garden entrusted to his care without falling into the trap of using solely a handful of personal favourites with all too obvious regularity.

In common with many eminent garden designers, Graham Thomas admires Gertrude Jekyll's use of 'drifts', narrow bands of one variety woven obliquely across the border. Drifts are undoubtedly pretty and by virtue of their narrowness allow plants on either side quickly to fill the space left by flowers with only a short season. However, they do not look effective if they run parallel to the border's edge and can create problems of relative heights in all but the widest borders, the leading edge of the drift standing proud of the surrounding planting while the rear edge risks being overwhelmed by taller neighbours. This could be a reason why, though so many garden designers profess admiration for drifts in principle, they are seldom used today.

There is no colour Graham Thomas dislikes, not even magenta, which Miss Jekyll so loathed; to him, it is 'a perfectly good colour. I don't understand why a human being should dislike any colour, except through association. I believe that the fundamental rule is that you don't have colours with yellow in them with colours with mauve and purple in them. If you put red at the top of the spectrum [wheel], you can have all colours on one side of it together or all colours on the other side together; they don't mix – that's the dividing line. After that, it doesn't matter what you do.' Though this is certainly a good general rule, I would like to see a planting in which small and

light-textured flowers of a bluish red from one side of the colour wheel were scattered over a ground of vermilion-red from the other, or vice versa, to shimmer with electric intensity. The Victorians loved to create similar shot silk effects by casting a haze of violet *Verbena rigida* (syn. *V. venosa*) over carmine or light crimson geraniums.

It is a curious fact that the colours to either side of pure red can seem so disharmonious. However, Graham Thomas feels it is perfectly acceptable to mix colours on either side of other primaries, such as violet-blue with greenish blue, because these are not as dominant as the reds, orange-red in particular; as with bluish red and orange-red, such a combination of blues seems to me disharmonious and to have the potential for either vibrancy or unpleasantness, though this is doubtless a matter of personal taste. He considers that 'yellows, as long as they are pale enough, go with almost any colour, though brassy yellows do not', while 'yellow is too

OPPOSITE Lavatera *'Barnsley' is a newcomer to the border, intended to provide some of the lightness the scheme demands. One of very few shrubby plants in this predominantly herbaceous border, it is cut back every spring; by early summer its racemes of palest shell pink appear and continue until the frosts. Biennial onopordums sow themselves in their allotted space at the back of the border, adding height and architectural form. Aconitum × cammarum 'Bicolor' produces its many spears of two-toned helmeted flowers from midsummer until the early autumn.*

BELOW *By late summer the borders have already taken on an autumnal air: seed heads of astilbe, tawny red* Persicaria amplexicaulis *(syn.* Polygonum amplexicaule*) 'Firetail' and tangled stems of* Crambe cordifolia *assume fawn and russet tones. The netting support is scarcely visible, though the hazel rods which hold it can just be seen behind* Aster × frikartii. *Magenta is represented by phlox and lythrum adding pizzazz while alabaster* Anemone × hybrida *'Honorine Jobert' and creamy* Artemisia lactiflora *leaven the whole.*

LEFT *In midsummer as rich purple delphiniums and* Salvia × superba *pass the peak of their display, the flat heads of* Achillea *'Gold Plate' (on the right of the plan) provide a classic contrast of colour and form in the hot-coloured border. Burnt orange* Helenium *'Wyndley' adds extra warmth in front while the bold glossy foliage of* Magnolia grandiflora *makes a splendid background.*

RIGHT, TOP *By late summer, the delphiniums and salvias show little colour and the border contains mostly flowers in the range yellow to red, leavened with white. Chosen before the staking method was adopted,* Salvia × superba *has proved too short to hide the vivid green mesh, regrettably not available in black or a more muted shade of green. A salvia whose foliage covers the mesh would probably now be preferable and this would also allow the flowers to be deadheaded, extending their season. The net is also visible through the sparse foliage of the achillea. There have been some changes since the border was designed, notably the addition of more white flowers:* Achillea ptarmica *'The Pearl' has replaced* Rudbeckia maxima, *a superlative glaucous-leaved coneflower which seldom achieves the magnificence of which it is capable unless cosseted in moist, rich soil. White phlox has also been added next to a carmine variety; this latter rather clashes with the orange crocosmias, tending too much towards blue-pink and scarcely the scarlet specified on plan. The few remaining lilies (bottom left), their leaves tattered by lily beetle, fail to supply the intended extra opulence.*

RIGHT, BOTTOM *Graham Thomas's original 1969 plan shows the bold scale of the planting, with groups typically 2.5×1.5m/8×5ft. Blocks slant in the same direction in an arrangement reminiscent of Gertrude Jekyll's drifts; this gives the planting a pleasing 'grain', more agreeable in appearance than groups sloping at random angles and much more attractive than ones parallel to the border's edge. However, the groups at Cliveden are thicker and more substantial than Jekyllian drifts, rendering them less susceptible to being crowded out by neighbouring plants. The plan also differs from Miss Jekyll's style in using a much wider range of plants.*

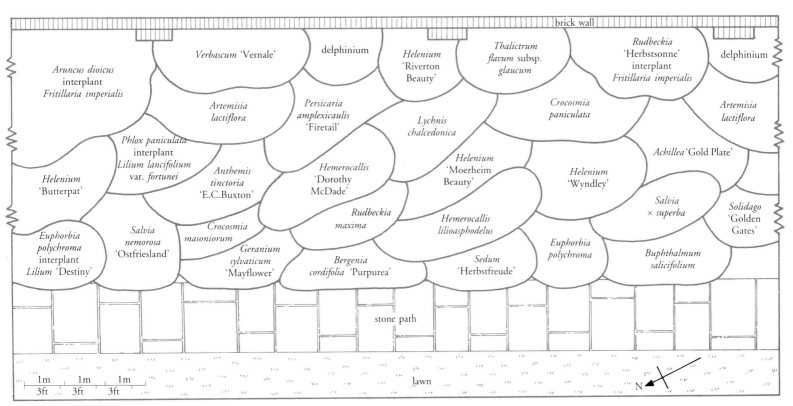

brick wall

Aruncus dioicus interplant *Fritillaria imperialis*

Verbascum 'Vernale'

delphinium

Helenium 'Riverton Beauty'

Thalictrum flavum subsp. *glaucum*

Rudbeckia 'Herbstsonne' interplant *Fritillaria imperialis*

delphinium

Artemisia lactiflora

Persicaria amplexicaulis 'Firetail'

Lychnis chalcedonica

Crocosmia paniculata

Artemisia lactiflora

Phlox paniculata interplant *Lilium lancifolium* var. *fortunei*

Anthemis tinctoria 'E.C.Buxton'

Hemerocallis 'Dorothy McDade'

Helenium 'Moerheim Beauty'

Helenium 'Wyndley'

Achillea 'Gold Plate'

Helenium 'Butterpat'

Euphorbia polychroma interplant *Lilium* 'Destiny'

Salvia nemorosa 'Ostfriesland'

Crocosmia masoniorum

Geranium sylvaticum 'Mayflower'

Rudbeckia maxima

Bergenia cordifolia 'Purpurea'

Hemerocallis lilioasphodelus

Sedum 'Herbstfreude'

Euphorbia polychroma

Salvia × *superba*

Buphthalmum salicifolium

Solidago 'Golden Gates'

stone path

1m 1m 1m
3ft 3ft 3ft

lawn

N

81

near to green to make a garden' on its own.

The colour scheme of the Cliveden borders is similar to that of the borders in the Pool Garden at Tintinhull in Somerset, designed by Phyllis Reiss in the 1930s, and so ably interpreted and enriched there until 1993 by Penelope Hobhouse. Here a border of soft colours to one side of the pool opposes one of hot colours on the other side, both borders containing white flowers to leaven the colour scheme and to give a unifying feature to otherwise disparate planting. Graham Thomas's design was not an attempt to repeat this scheme but was chosen because he felt that either a complete mixture or a pair of borders in the same restricted colour range would be too dull. Unlike Mrs Reiss, he dislikes the combination of stark white with hot colours and for that reason chose to leaven the scarlets, oranges and golds with creamy white. *Phlox paniculata* 'Mia Ruys' was chosen instead of 'Frau Antoine Buchner'

because its yellow eye gives it a creamy tone that is much less harsh than pure white.

The soft-coloured border has always given problems. Graham Thomas says, 'they are receding colours so it's important to add plenty of light pink and white; mauves don't show up at all, particularly when you're looking at the borders from some distance.' Shasta daisies, required to give the essential white, resolutely refused to grow and although the addition of *Achillea ptarmica* 'The Pearl' and *Lavatera* 'Barnsley' has helped, there is still a lack of these paler tones.

The red brick wall at the back of the border flatters these soft colours less than the hot colours opposite. Pinks that tend towards lilac, such as *Lavatera* 'Rosea', can look disagreeable. Some stretches of wall have yet to be covered by clematis or suitable evergreens to avoid such clashes. However, as Head Gardener Philip Cotton points out, the 72ha/180 acres of Cliveden do not leave time

to train clematis up the wall as meticulously as at Sissinghurst, whose smaller scale allows much more labour-intensive methods.

For the same reason, staking must be achieved simply and there is no question of dealing with each group individually using brushwood as at Sissinghurst, nor is there time in winter to gather the vast number of

BELOW, LEFT *For a brief moment as spring passes into summer, yellow* Hemerocallis lilioasphodelus *and blue geraniums dominate both borders. Creamy* Aruncus dioicus *adds height and lightens the scheme. The panels of netting which will support the border are in place though not yet hidden by foliage.*

BELOW *A detail shows how panels of netting are fixed at the front of the border. Hazel rods are woven through the edge of the nets and fixed to sturdy angle irons, keeping the netting taut. The foliage of* Hemerocallis lilioasphodelus *never quite hides the netting, though the plume poppies engulf it completely.*

peasticks needed. Furthermore, Graham Thomas hates to see stakes looking 'like Burnham Wood' and considers netting to be preferable by far. This system of staking was devised at Crathes Castle in Scotland using 15cm/6in mesh black netting put in place horizontally before the border grows. The much shorter Scottish growing season ensures that the netting very quickly disappears; this rapid growth also gives little opportunity for weeds to compete.

The system was introduced at Cliveden in 1987 but not without some difficulties. The borders are much wider than at Crathes and so the netting had to be set in two ranks of panels, each about 2m/7ft wide, sloping up towards the back. Stout angle irons keep the corners of the panels in place and are left in situ throughout the year, even in winter. Hazel rods hold the outer edges of the nets, keeping them taut.

Black netting was no longer available and the shade of green plastic now used for pea nets is especially vivid and scarcely less obtrusive than purple or orange, a particular annoyance where plants are too short to hide the net. This problem could be overcome by replacing with taller varieties; although we might not like the evenly banked appearance this would give, it is still possible to have tall plants coming to the front of the border to break the monotony.

At Cliveden a longer season allows the weeds to grow up with the plants; the nets hamper weeding, though gaps around the panels give access for periodic assaults. Philip Cotton finds that the task of trying to disentangle the netting from the haulm of the plants is just not worth the effort; when the time comes to cut the plants down in early winter, the whole lot, net and haulm together, is discarded. Inevitably some plants have already started to die back by midsummer but the parchment colour of the dead stems is not unattractive and their skeletons continue to give bulk and structure

to the borders through the latter part of the season. Overall, Philip Cotton feels that the cost of the netting is good value and that the system saves a great deal of time.

Bindweed is the worst of the borders' weed problems. Minor infestations are dealt with by pulling it out by hand but translocated herbicides are used if the invasion threatens to become overwhelming. Periodic emptying, sterilizing and replanting of the borders every ten years or so also gets rid of most weeds, although some of the bindweed usually manages to struggle back from beneath the treated soil.

The borders' main problem is that they are on gravel overlying chalk which dries out very quickly, making it almost impossible to succeed with some of the plants on Graham Thomas's original design, particularly monardas and phlox. Mulching with well-rotted manure helps but the borders still become so dry from time to time that less tolerant plants fail. Many gardeners, worried by the increasing frequency of dry summers, have considered irrigation and at Cliveden this possibility was investigated. However, local mains water pressure is too low, and a system using a large tank to collect rainwater and a pump proved too costly to install.

The display at Cliveden starts with tulips in late spring, anticipating the summer colour scheme of reds and yellows on the hot side with pinks and bluey tones opposite. The tulips are not lifted after flowering but are left in situ, to be topped up from year to year if numbers dwindle. Within a few weeks, yellows and blues predominate on both sides, *Hemerocallis lilioasphodelus* (syn. *H. flava*) and *Geranium* 'Johnson's Blue' playing the leading roles. By early summer, the vertical accents of delphiniums dominate both sides with spires of yellow verbascum on the hot side for contrast. However, there have been losses among delphiniums on the opposite side; replacement plants are grown in the nursery until they are big enough to fend for

themselves among the other giants at the back of the border.

Lilies were interplanted with several of the groups on both sides but have not succeeded because of the ravages of drought and lily beetle, a pestilence which has spread slowly but inexorably beyond its original epicentre in Surrey. Few plants can add such an air of sumptuous quality as lilies, particularly if used in sufficient quantity. Graham Thomas remembers the good old days when 'one used to go to an auction sale in London and buy a crate of *Lilium longiflorum* for about sixpence; that's why the Jekylls of the time used them so much – they were only used once and then thrown away'.

Graham Thomas is philosophical about the need to adjust planting plans and the desirability of accommodating good new varieties. If something does not thrive, one thinks of something else. A feature he admired about Mrs Reiss's Tintinhull borders was the way she would add new plants, both to improve on what was there and to give variety from year to year, constantly revitalizing the planting. He owns that 'nobody can make a border absolutely perfect on paper. It's not easy. Borders require some adjustment; we grow plants from all over the world and expect them all to do well in one border.'

The loss of phlox and monarda has resulted in a lack of certain colours, particularly bright reds and oranges on the hot-coloured side. *Penstemon* 'Andenken an Friedrich Hahn' (*P.* 'Garnet') has proved a successful substitution for one such loss among the soft colours; it may be that scarlet penstemons would serve to replace the likes of *Monarda* 'Cambridge Scarlet'. Philip Cotton intends to stick to Graham Thomas's scheme as closely as possible, using different plants to achieve a similar effect where the originals have failed.

Every summer he makes a note of any gaps in the borders or of groups that have become too large and are encroaching on their

neighbours so that the planting can be adjusted. The plants are mostly chosen not to need regular replanting so one important element of the labour requirement is greatly reduced. A few plants are bought in every year to fill gaps; however, plants usually need to be grown on in the nursery for a year to be big enough to avoid being lost.

Onopordums, a later introduction suggested by Graham Thomas about eight years after the borders were first planted, are allowed to self-sow. Occasionally they need to be transplanted when young to get them in the right place.

The scale of the plant groups at Cliveden is so large that one is not aware of the number of plants in each. However, in smaller-scaled schemes, as at Powis Castle in Wales, Graham Thomas's planting plans show that he is no believer in the old rule that plants should be grouped in odd numbers of one, three, five or seven. This he considers to be 'a lot of silly nonsense': whereas three plants cannot fail to be seen as a triangle, the addition of a fourth plant inevitably makes a more interesting shape. There is often no need for a fifth plant. In narrower borders, his plans often feature groups of just two plants, though a third will generally be placed within the distance of a metre/yard.

A firm believer in mixed borders, Graham

In early summer, only a small proportion of the flowers are in bloom. The strong verticals of delphiniums, Verbascum *'Vernale',* Salvia × superba *and crocosmia leaves, plus the bold colour contrast of purple and yellow, are sufficient to carry the design amid expanses of less structured foliage and a haze of* Crambe cordifolia. *Spikes of goldenrod assume a pleasing shade of chartreuse as they approach flowering while towards the front yellow* Buphthalmum salicifolium *is a stalwart performer. Though the borders were designed to be viewed at right angles, the fact that each uses a single colour scheme throughout its length helps ensure that lengthwise and oblique views such as this are equally attractive.*

Thomas does not like ordinary herbaceous borders and thinks that they generally lack richness. However, at Cliveden he felt an herbaceous border was more appropriate because the plants used to be required to perform only during the summer months. For the amateur's garden, he believes the admixture of shrubs is almost essential to give bulk and continuity throughout the year.

John Sales feels that the borders are seen to their best advantage obliquely or along their length. Accordingly, mulberry trees have been planted in the lawns on either side of the entrance drive to cut off the views at right angles, while their staggered planting will leave gaps for the oblique views. Visitors have been further encouraged to see the borders along their length by opening an entrance in the yew hedges at the end of each York stone edging path.

Such a change in concept calls into question several of the design features of the borders: when the mulberries have grown and the views at right angles have been blocked, will the large plant groups be correctly scaled or should they be reduced for more intimate viewing? Are the openings in the hedge sufficiently architectural for what have become important terminations to the lengthwise views along the borders? Now that the stone edging has assumed more of the character of a path, should it be widened?

Throughout the phased replanting that Philip Cotton plans to complete over the coming years, all details of the borders' design and planting will be reconsidered. The optimum balance between Graham Thomas's original concept and the present way the borders are viewed will be sought; plants will be chosen to replace those that have failed, remaining faithful to the skilled interplay of textures and colours originally devised. The borders will remain a tribute to this most eminent artist-plantsman whose life's work has been a source of inspiration to gardeners world wide.

A RAINBOW BORDER

The Priory, Kemerton

The long Top Border at the Priory, Kemerton in Hereford and Worcester, created by Peter and Elizabeth Healing for late summer and early autumn display, is justly renowned as a dazzling example of a colour-schemed border. From pink and white at the end nearest the house, the colours change over quite a short distance to yellow contrasted with blue, moving through orange to a triumphant central section of richest red. The colour scheme then reverses, the pinks giving way to a white border at the furthest end, separated from the main length of the Top Border by a pergola across a grass path. Since Peter Healing's death, his widow has continued to maintain the garden and open it to the public, with help from Charles Brazier, gardener for forty years until his retirement in 1993, and his successor Stephen Hocking.

Mrs Healing's childhood home had a grand herbaceous border, designed by her mother with help from the formidable Ellen Ann Willmott, a renowned plantswoman, garden designer and author of the sumptuous work *The Genus Rosa*. The considerable significance of Miss Willmott's designs has yet to be assessed by garden historians, though Elizabeth Healing considers they lacked the artistic flair of those of Gertrude Jekyll, who had been closely involved with the Arts and Crafts movement and was accomplished in so many arts and crafts herself. Peter Healing's introduction to horticulture came later in life through his mother-in-law's gift of a copy of William Robinson's classic work, *The English Flower Garden*, which accompanied him during service in World War II.

When the Healings moved to Kemerton in 1938, the border had no colour scheme: their predecessors had planted it higgledy piggledy with delphiniums, lupins, gladioli and miscellaneous herbaceous plants. A yew hedge was planted behind the border, which Elizabeth Healing always intended should provide a background to the planting;

The Top Border in late summer. From a separate white section, dominated here by Artemisia lactiflora *'Guizhou', the colours change slowly and subtly through pink to violet and purple. Then comes a sudden transition to orange and yellow, followed by a gradual progress through red, whereon the colour scheme reverses, bright yellow being set against lavender-blue and purple. Passing through pink to*

white, the border ends in a mist of gypsophila. Seen along its length, the two more rapid changes seem so compressed that the border appears to consist of two blocks of soft colours separated by one of yellow, orange and red.

Diagram labels:

yew hedge

grass path

1m 3ft   1m 3ft   1m 3ft

Clematis 'Perle d'Azur' · Ceanothus 'Gloire de Versailles' · Clematis 'Royal Velours' · Rudbeckia 'Herbstsonne' · Rosa glauca · Clematis 'Kermesina' · Rosa 'Geranium' · Echinops ritro · Helianthus 'Capenoch Star' · rudbeckia double yellow · Atriplex hortensis var. rubra · Atriplex hortensis rubra · Helenium 'Wyndley' · Achillea 'Gold Plate' · Kniphofia uvaria 'Nobilis' · Crocosmia 'Lucifer' · canna · A · Lavatera 'Barnsley' · monarda · Helenium 'Moerheim Beauty' · Rosa 'Golden Years' · S · R · Dahlia 'Bishop of Llandaff' · Monarda 'Mrs Pe...' · Aster × frikartii 'Mönch' · Kniphofia 'Green Jade' · Solidago 'Strahlenkrone' · Crocosmia 'Star of the East' · Phlox paniculata 'Starfire' · G · Phygelius capensis · A · G · C · Malva alcea · Salvia × superba · Heliotropium 'Chatsworth' and H. 'Marine' · Helenium Sunshine hybrids · Rudbeckia fulgida deamii · Helenium Sunshine hybrids · antirrhinum red · Verbena 'Lawrence Johnston' · P · A · P · F · F

ABOVE *The castor oil plant* Ricinus communis *'Gibsonii' adds an exotic, almost tropical, note to the central section of the border, its rich tones harmonizing with vermilion* Crocosmia *'Lucifer' and coral red* Monarda *'Mrs Perry'. The monarda, softer in colour, shorter and more even in height than 'Cambridge Scarlet', is an outstanding performer every year yet unaccountably seldom seen in gardens.*

ABOVE, RIGHT *This part of the planting plan shows how, from the central red section of the border, the colours shift to orange and yellow. The abrupt transition to lavender-blue and purple gives some striking contrasts before the tones change to more peaceable pink. The hot colours and bold foliage of castor oil plant, cannas and crocosmias give the section an exotic feel altogether different from the softer-coloured sections.*

however, her husband could not reconcile himself to the idea, thinking the yews robbed all goodness from the border. His solution was to put a grass walk at the back of the border distancing the yews from the plants, but he remained sceptical about the benefits of a tall hedge as a background. Now at last the hedge has reached the height of 1.8m/6ft that Elizabeth Healing originally envisaged. Not only does it provide a perfect foil for the border but it gives enclosure to a June garden hidden on its other side.

It is curious that most British gardeners feel uncomfortable with a backless border, though in North America these are not uncommon, sometimes paired with a path in between. The insistence on enclosure, creating the feeling of an outdoor room, seems to be one of the most characteristic

features of what is called the English garden (though it might more accurately be called British). The double borders created by Graham Thomas in the centre of the Rose Garden at Mottisfont Abbey in Hampshire are a rare exception, yet perhaps they work so well because they make the visitor flee from the open space between the borders into the secluded rose-lined walks which are the garden's *raison d'être*.

After the war, Peter Healing filled the border with his prize dahlias, to the dismay of his wife who thought them dreadful. With a little persuasion from her, he was encouraged to abandon dahlias for orchids, leaving the borders free for a different approach, though a few of the choicer dahlias were kept. Elizabeth Healing suggested herbaceous borders in 1957 and her husband liked the

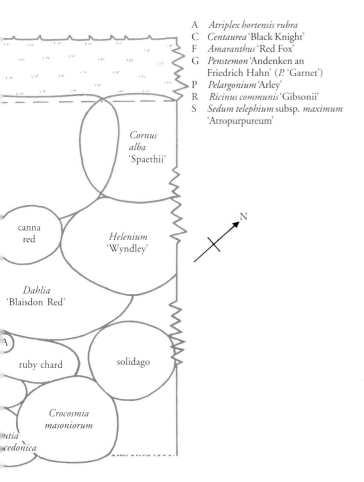

A   *Atriplex hortensis rubra*
C   *Centaurea* 'Black Knight'
F   *Amaranthus* 'Red Fox'
G   *Penstemon* 'Andenken an
     Friedrich Hahn' (*P.* 'Garnet')
P   *Pelargonium* 'Arley'
R   *Ricinus communis* 'Gibsonii'
S   *Sedum telephium* subsp. *maximum*
     'Atropurpureum'

Cornus alba 'Spaethii'

canna red

Helenium 'Wyndley'

N

Dahlia 'Blaisdon Red'

A

ruby chard

solidago

Crocosmia masoniorum

...utia ...cedonica

idea. By 1959, the borders were beginning to take shape and to demonstrate grouping and colour scheming. Though Elizabeth Healing was by then occupied with a family, important decisions about the style and planting of the garden, including the graded colour scheme, were taken together.

Peter Healing always considered himself a plantsman rather than a garden designer and collected choice plants during his travels around the country. Elizabeth Healing remembers what a slow job it was and how empty the borders seemed to be in those early days. Gertrude Jekyll's *Colour Schemes for the Flower Garden* was a source of inspiration, though once the Healings had embarked upon their task of a colour-schemed border, they no longer looked back to Jekyll: the task 'went on from strength to strength, it had *got*

us and we really didn't look to see what other people were doing', Elizabeth Healing remembers. 'We planted perennials together in small groups to begin with to get the colours right and the infilling was done with annuals. That went on for years until gradually we realized we'd got it.' The effect they had wanted had been achieved and most of the gaps in the border had disappeared.

Peter Healing had disliked borders for early summer that include flowers like lupins and delphiniums because later in the season they leave gaps demanding to be filled with both costly and labour-intensive bedding. He decided that it was much better to concentrate on a period of two to three months from midsummer onwards. For spring, Mrs Healing has added tulips that match the colour scheme as it will develop

ABOVE *The tranquillity of pink, white and lavender comes as welcome relief from the excitement of the border's hot colours. The varied flower shapes of monarda, echinacea, dahlias,* Penstemon *'Alice Hindley' and* Lavatera *'Barnsley' blend harmoniously.*

OVERLEAF *A progression of plant groups along the border shows changing colour schemes over a short distance.* (LEFT) Ceanothus *'Gloire de Versailles',* Clematis *'Perle d'Azur', monarda,* Salvia sclarea *var.* turkestanica, *echinops and* Malva sylvestris *var.* mauritiana *shimmer in the sun.* (CENTRE) *Viticella clematis 'Royal Velours' and the same bush of ceanothus are contrasted on its other side with* Achillea *'Gold Plate'.* (RIGHT) *Bomb blasts of rich green crocosmia leaves topped with flaming flowers are complemented by rich* Clematis *'Kermesina', dusky red orach and* Monarda *'Mrs Perry'.*

later on, yellow with yellow, red with red and white at the end of the White Border. Wallflowers were tried but abandoned because they looked untidy and had to be replaced with summer flowers before they were quite over.

The rarity of such ambitious borders in late-twentieth-century gardens is doubtless due in part to the considerable cost of the large numbers of plants needed. Mrs Healing is in no doubt that the Kemerton borders would not have been possible without her husband's enthusiasm for propagation. In latter years, a mist unit has facilitated the speedy propagation of all the plants needed for the garden, with an ample surplus available for sale to visitors, the very best souvenir of the garden and its choice range of plants.

In spite of Mrs Healing's determination to minimize labour, she considers some annuals indispensable. They are all raised in a propagating house, maintained at a minimum temperature of about 7°C/45°F, from sowings at the beginning of spring (during March), starting with ornamental beetroot and ruby chard, followed by the cornflowers, antirrhinums, amaranthus and cosmos. Finer seeds are sown on the surface in trays containing a mixture of vermiculite and peat, larger seeds being covered with horticultural grit. The trays are covered with clingfilm which causes less condensation than other

Red dominates the centre of the border in a scheme reminiscent of the Red Borders at Hidcote but made brighter by touches of orange and plentiful gold-variegated and green foliage. At both Hidcote and Kermeton, Rosa 'Geranium' is a star performer, its graceful arching branches giving height. Self-sown purple orach weaves itself through the border, its leaf colour repeated by that of Dahlia *'Bishop of Llandaff'. 'Blaisdon Red', among the most sumptuous of blood-scarlet dahlias, supplies the necessary intensity of colour to carry the scheme.*

transparent coverings; if polythene or glass is used, drops of condensation can fall back onto the surface of the trays and cause damping off. Black polythene is used over the clingfilm to prevent the trays overheating in sunlight. When the seeds germinate, the covering is removed and the young seedlings are left for a few days to harden off before being pricked out into trays. Young plants are moved into individual pots before being planted out in late spring once the danger of frost has passed. Castor oil plants are grown from seed from the previous year's plants, sown individually in small pots and later potted on to a larger size.

Only a few tender plants are grown that have to be taken in for the winter. Cannas and dahlias are the most important of these, penstemons generally being sufficiently hardy to survive without protection. This is partly because the tougher sorts are chosen: 'Andenken an Friedrich Hahn' ('Garnet'), 'Schoenholzeri' (syn. 'Firebird') and 'Rich Ruby', grown mainly for the Red Border nearer the house but occasionally used in the Top Border, have all proved themselves over a number of typically Zone 8 winters, though 'Mother of Pearl' and 'White Bedder' (syn. 'Snow Storm') are a little less tolerant of frost. However, penstemons are not long-lived: regular raising from cuttings ensures greater vigour and more prolific flowering, as well as providing an insurance against unusually severe winters. Elizabeth Healing finds that penstemon cuttings taken in spring or in mid autumn stubbornly refuse to flower; accordingly they are propagated only from midsummer to early autumn, new plants being overwintered under cover. Earlier propagations are potted singly and held in an unheated polythene tunnel but later ones remain in trays of twenty young plants, overwintered in cold frames protected from the severest frosts by tarpaulins.

Autumn is perhaps the most important working season in the Kemerton borders.

Once early frosts have thoroughly blackened the dahlias, their tubers are lifted and labelled, then turned upside down to dry off for two to three weeks in a frost-free environment before being placed in long trays covered with dry soil, the right way up again, and kept beneath the staging of a cool but frost-free glasshouse. The soil is devoid of nutrients and is merely a medium to prevent the tubers from drying out excessively. When shoots start to appear in the spring, the tubers are placed in trays of better soil in the light. If only a few new plants are needed, the rootstocks may be divided but if more are required, new shoots may be used as cuttings. The dahlias are planted in the borders once the danger of frost has passed in late spring. Cannas are lifted at the same time as the dahlias and planted in soil in wooden orange boxes, also kept beneath the staging until new shoots start to appear.

After removing cannas and dahlias, the border plants are cut back. Elizabeth Healing finds it difficult to attempt replanting in the spring when the borders look bare and it becomes hard to visualize what the colours were; changes to the planting are made in autumn while memories of any shortcomings are still clear in the mind. Overlarge clumps are reduced and there is a constant attempt to find perennials that will provide as bright and long-lasting an effect as the remaining patches of annuals.

Mushroom compost used to be dug through the border in autumn but the lack of a mulch to suppress weeds proved to be a problem, the weeds growing at the very time when gardeners are most preoccupied with other work. The new system at Kemerton is to fork garden compost through the border in autumn and apply a mulch of mushroom compost over the top in spring. Then there will be no need to disturb the soil, bringing weed seeds to the surface, and the mulch should be thick enough to prevent seedling weeds from pushing through.

with which the border begins seem well separated from the hot colours, and the purples and blues can be seen harmonized with pinks or contrasted against yellow. The transition from yellow through orange to red and back again is achieved much more gradually so that the central section of the border has a more spacious and relaxed feel. At close range, the lengthwise view along the border throws contrasting colours into startling juxtaposition; some might find this refreshing, invigorating, even dazzling, but there will undoubtedly be others who find the colour contrasts uncomfortable and hard to take.

Foliage textures are especially pleasing in the central section of the border. Here is a tapestry of exotic-leaved cannas and castor oil plants contrasted with sword-like crocosmias and red hot pokers while self-sown purple orach threads itself throughout, providing a unifying note. The spires of *Kniphofia uvaria* 'Nobilis' find an echo in the columns of Irish juniper and fastigiate golden yew appearing above the top of the hedge at the back of the border. The very choicest of Peter Healing's collection of dahlias still grow in the border. 'Bishop of Llandaff' with scarlet flowers and finely divided dark foliage survived here when it had all but disappeared from other British gardens; it is now perhaps the most widely grown of all dahlias in the British Isles. 'Blaisdon Red' is a particularly sumptuous Decorative sort, scarlet shaded with deepest blood-red, while 'Grenadier' has dark foliage, less divided than the Bishop's, and semi-double flowers.

The border is deep enough to accommodate the largest shrubs. Notable among them are feathery-leaved *Caragana arborescens* 'Lorbergii' into which weaves soft-blue-flowered *Clematis* 'Perle d'Azur'; a large bush of *Rosa* 'Geranium' some forty years old, its branches dripping with hips of sealing-wax red, arches over brilliant gold-variegated *Cornus alba* 'Spaethii'. Further forward in the border, *Lavatera*

Bamboo canes are used for staking, painted green to make them less conspicuous. The tone of green is carefully chosen, dark and subdued so that any canes that are visible attract as little attention as possible. Generally the canes are hidden inside the clump they support, with garden twine looped from cane to cane around the perimeter of the plant, tracing the outline of a trefoil or quatrefoil.

Concealment within the plants is a great advantage: even when painted green, the canes' abruptly ended straightness looks unnatural and draws the eye as effectively as an exclamation mark.

Elizabeth Healing considers that the border is best seen at right angles from a little distance rather than at point blank range or along its length. Thus the pinks and whites

'Barnsley' has proved itself a stalwart performer, used among pink flowers at both ends and flowering prolifically throughout the summer until the frosts.

The Top Border at the Priory is outstanding for the richness of its colours, the length of its display and its harmonies and contrasts of foliage texture. It needs to be at its peak for a shorter period than the West Court Borders at Hardwick, which must remain in prime condition from earliest midsummer until the frosts. The Kemerton border uses plants that are, on the whole, not so labour intensive: the dahlias which contribute significantly to the display are relatively good value, needing no cosseting through the winter when they are dormant and each growing to make a plant of considerable size. Furthermore, the Priory's

border is deeper and more spacious than Hardwick's, allowing more complex counterpoints and contrasts between similar and dissimilar plants.

The best borders are not the result of a day or two devising a plan, followed by a season or so in which the plan is realized: they are the result of years of hard work and self-criticism, constantly questioning and refining plant associations and horticultural techniques, so that little by little every aspect of the border approaches the ideal. The dedication, professionalism and industry applied by Peter and Elizabeth Healing to their garden for so long ensured that the borders at the Priory attained a rare excellence, to become an inspiration to all gardeners and a tribute to their lifetimes' work.

ABOVE *A vine-covered pergola divides the pink section of the border from a separate bed planted with white alone. Viticella clematis 'Alba Luxurians' is mirrored by shell-pink* Lavatera *'Barnsley', with dahlias, mallow and monarda beneath in richer rose.*

OPPOSITE *An oblique view of the border from a short distance away shows an intensely vivid transition in colours. Here, in midsummer,* Phlox paniculata *'Starfire' dominates the foreground. The mound of solidago marks the transition from vibrant colours to the purple-blue of* Salvia × superba *and monarda beyond. The coarsely handsome* Malva sylvestris *var.* mauritiana *stands pre-eminent among the purples. Some gardeners will find this transition exciting and exhilarating, though others might find it strident. Elizabeth Healing likes views of the border from a short distance away but prefers views at right angles from front to back.*

Bold Colour in a Mixed Border

Great Dixter

Colour combinations in Christopher Lloyd's Long Border at Great Dixter in East Sussex are blatantly unconventional. Some people would consider they cross the threshold from the ecstatically pleasing to the exquisitely painful. But Christopher Lloyd is used to arousing such controversy. Many gardeners cannot agree with his views on gardening. 'He's so opinionated!' they say; in other words, their own views differ. For the majority, however, it is just these opinions that make his gardening exciting and his writing entertaining, always challenging and wittily and elegantly expressed. It does not matter if we do not agree: he does not really want us to, but merely tries to make us question our prejudices and be more daring in the way we plant.

Trained in horticulture at Wye College, Christopher Lloyd sees the art and technique of gardening through the eyes of a professional, a rare approach among private garden owners. At Great Dixter his planting demonstrates that he practises what he preaches and the border shows much of his individual style.

Although elements of the garden were designed by Sir Edwin Lutyens, his frequent associate Gertrude Jekyll was not asked to provide planting plans and the Long Border planting was first devised by Sir George Thorold. Miss Jekyll did, however, bless the seven-year-old Christopher, an event that seems to have had singularly little effect on his development as a gardener: there is scarcely a trace at Dixter of the Jekyll style; all we see is pure Lloyd.

Carefully controlled colours shading into one another and evenly graded heights, the floral catafalque style we sometimes see at the

In the view along the border away from the house, towards the Lutyens-designed bench, the juxtaposition of colours is typically daring, with golden Inula magnifica, *scarlet rose 'Florence Mary Morse' and mauve* Phlox maculata *'Alpha' in the foreground. Beyond, bright magenta cranesbills, shocking pink phlox and billowing clouds of smoke bush contribute to the dramatic effect with a haze of purple* Verbena bonariensis *shimmering against sombre yew. The broad path provides a perfect foil for riotous colour, allowing plants to tumble forwards to be inspected at close range.*

Chelsea Flower Show, are out. Heights are uneven with some tall plants coming right to the front and deep canyons running to the back of the border, allowing us the rare opportunity of seeing some of the bigger plants furnished with fine foliage from top to toe.

Colour combinations are usually daring. Perhaps not everyone will believe Christopher Lloyd when he says he could produce a pink and yellow border without any unpleasant colour combinations; I do, and long to see it. He admits, 'I haven't got the inhibitions about colour that most people do – I like all the colours – but I wouldn't plant a pink 'Zéphirine Drouhin' rose next to a scarlet *Lychnis chalcedonica*; there are limits.'

Magenta, anathema to Miss Jekyll, is a favourite at Dixter and there are few flowers in wishy-washy or subfusc tones. Strong yellows and reds are much in evidence: certainly there is a wish to show the usefulness of rich and vibrant colour but one suspects also an impish desire to alarm excessively aesthetic sensibilities.

On planting in a restricted colour range,

Christopher Lloyd says, 'I think people really need hand holds when they start gardening and they find it easier to limit themselves in various ways. Usually they'll say, "Oh, I can't bear orange" or "I can't bear yellow" or "If I can bear yellow, I must have all my yellows together." I can quite understand that because there is a tremendous choice of plants and it is very daunting for people to know where to start; but having gone through that and having been doing it all my life, I don't find it daunting at all. I like to use all the yellows in different ways, not only all the colours but all the different plants, every sort of plant. If I had a spare cactus I'd find a place for it, or an agave.'

Plant associations are worked out by considering what would combine well with a cardinal plant. Thus yellow Mount Etna broom (*Genista aetnensis*) has *Inula magnifica* planted beneath with flowers of the same hue, a colour harmony but a contrast of texture. In other cases it is the colours that are contrasted. The effect is unintentionally episodic, not a Bruckner symphony in plants

conceived as a monumental unified whole but a series of incidents. This is perhaps why Christopher Lloyd so enjoys the view at right angles to the border, showing clearly each individual association, rather than obliquely along the whole, when some of the key plants in each group disappear behind their neighbours and colours that are deliberately placed away from each other seem startlingly (or, perhaps worse, ineffectively) juxtaposed.

Great Dixter is not a garden for the tidy-minded. Stems of tawny tattered cardoons, teasels and most herbaceous plants are allowed to wither and are enjoyed for their stately but shabby beauty until the border is cut down and worked through in the spring. Round heads of *Allium christophii* detach themselves and trundle along the border like tumbleweed scattering seed as they go; the alliums and many of the other plants sow themselves freely, often in great drifts with odd outlying plants. Many seedlings are poorly placed or too crowded and need to be rooted out but Christopher Lloyd enjoys their informal effect, especially since some manage

Today's planting, shown here at the far end of the border, provides an interesting comparison with Sir George Thorold's original plan. Some similarities include a proportion of shrubs such as lilacs and purple hazel and plants like Miscanthus sinensis *'Zebrinus', grown for contrast of texture. Thorold's plan grouped together like plants (asters or pokers or delphiniums) and separated broad bands of rich colour with white flowers. Although there were some classic colour contrasts, there was no attempt at the sort of colour progression favoured by Miss Jekyll. The differences between Sir George's design and the present border are telling: the original had heights carefully graded so that no tall plant was found among shorter neighbours, a treatment Christopher Lloyd detests; now the sizes of plant groups have become much more diverse, from small individual plants to broad swathes of cardoon 3m/10ft across; and there are far fewer white or pastel-coloured flowers. It is hard to imagine Christopher Lloyd ever setting down such a scheme on plan, a system which cannot allow the informality nor the constant reworking and experiment which are essential features of the Long Border today.*

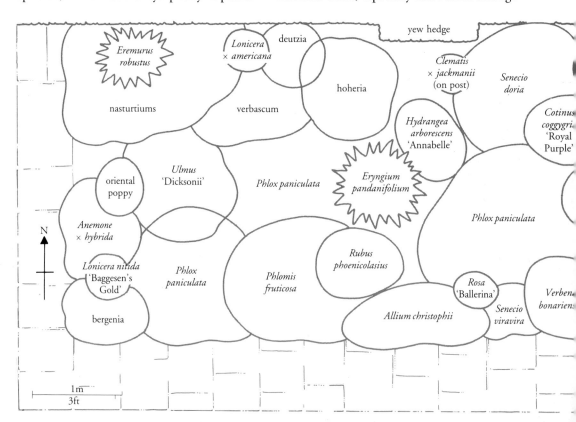

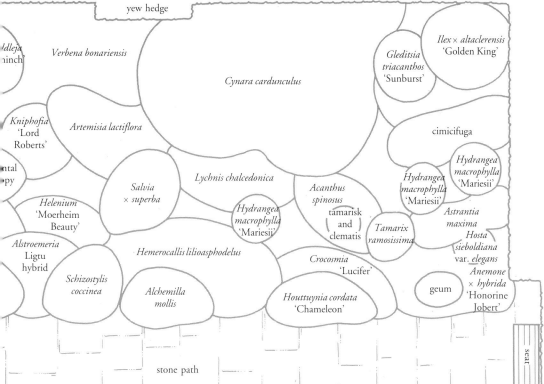

yew hedge

Verbena bonariensis

Ilex × altaclerensis 'Golden King'

Gleditsia triacanthos 'Sunburst'

Cynara cardunculus

dleja inch

Kniphofia 'Lord Roberts'

Artemisia lactiflora

cimicifuga

Hydrangea macrophylla 'Mariesii'

ntal py

Lychnis chalcedonica

Acanthus spinosus

Hydrangea macrophylla 'Mariesii'

Salvia × superba

Astrantia maxima

Helenium 'Moerheim Beauty'

Hydrangea macrophylla 'Mariesii'

tamarisk and clematis

Tamarix ramosissima

Hosta sieboldiana var. elegans

Alstroemeria Ligtu hybrid

Hemerocallis lilioasphodelus

Crocosmia 'Lucifer'

Anemone × hybrida 'Honorine Jobert'

Schizostylis coccinea

Alchemilla mollis

Houttuynia cordata 'Chameleon'

geum

seat

stone path

ABOVE *An oblique view across the centre of the section shown on plan, from* Phlomis fruticosa *to* Cynara cardunculus, *shows a wealth of form and texture: stiff candelabra of verbascum, bold cardoons, feathery tamarisk and gleditsia provide a setting for sizzling colour. Here pinks with blue (phlox) are mixed with reds and pinks with orange (*Helenium *'Moerheim Beauty',* Kniphofia *'Lord Roberts',* Alstroemeria Ligtu hybrid*), a combination many gardeners would find shocking.*

OVERLEAF *The shabby beauty of seeding stems is enjoyed to the full at Great Dixter. By midsummer the flowers of* Allium christophii (RIGHT) *have become a galaxy of jade-green pods tipped with parchment spines, borne on a starburst of stems that appear even pinker among the froth of smoke bush. In autumn, the tawny heads of cardoons* (CENTRE), *at the furthest end of the border, provide a satisfying interplay of texture and form with feathery* Gleditsia triacanthos *'Sunburst' and* Ilex × altaclerensis *'Golden King'. Teasels and cardoons* (LEFT) *tone well with vibrant* Salvia uliginosa *and glaucous eucalyptus.*

99

to strike up surprisingly successful associations with other plants, as good as anything the gardener could contrive. Miss Jekyll would not have approved of leaving so much to chance – but then her clients did not have a Lloyd to decide which seedlings were well sited and which were not. The arrival of *Verbascum olympicum* in front of salmon-pink *Alstroemeria* Ligtu hybrids was just such a serendipitous combination, enjoyed all the more for its daring colours.

Since the border was first planted there has never been any attempt to sweep everything away and start afresh. Christopher Lloyd values the process of building on what is already there, strengthening the existing structure and fleshing out the bones (the shrubs, some of which were there from the start) with bright-flowering herbaceous plants, annuals and good foliage.

Like most modern borders, the Long Border is designed to be attractive all year. Unlike many of Miss Jekyll's clients, Christopher Lloyd does not go away to shoot grouse each August, nor does he spend his winters in the south of France. The shrubs give much of the winter effect, particularly evergreens such as the fine *Ilex × altaclerensis* 'Golden King' at the upper end of the border, planted in 1954 and now his favourite plant, berrying well and providing year-round interest. This is typical of the shrubs he likes, slow growing and with masses of character: 'the slower they grow, the better they are in the long run'. Other favourites are a gnarled *Phlomis fruticosa*, an unusual wavy-leaved form planted in 1949 which in spite of its gaps is full of charm, and a large and spectacular 35-year-old *Euonymus fortunei* 'Silver Queen'. These and the sere stems of herbaceous plants keep the border looking full of interest even in midwinter, the rich green yew hedge behind providing the perfect foil.

Many of the shrubs, and even some small trees such as eucalyptus and *Salix alba* var. *sericea*, are pruned hard each spring to

produce bright new foliage. The willow has parasitic toothwort, *Lathraea clandestina*, growing on its roots for spring flower and golden elm behind for vivid summer contrast. *Gleditsia triacanthos* 'Sunburst' is also treated as a large shrub and pruned to produce healthy young leaves. Even in October the new foliage is fresh bright yellow, a contrast with the older green leaves; as with the genista/inula combination, it provides a colour harmony with the 'Golden King' holly but a complete contrast in form – light and feathery against the solidity of its neighbour.

In spring the border is cut down, most of the shrubs pruned and a proportion of the herbaceous plants divided; some plants, for instance phlox, are split every three years, others such as *Eryngium pandanifolium* and *Salvia × superba* only once every ten years, Japanese anemones, pokers and alstroemerias not at all. By this time most of the spring bulbs have started to push through and can be avoided. A mulch of mushroom compost follows and when the tulips begin their display most of the important routine work has been done. A heavy dressing (100-130g per m²/3-4oz per sq yd) of general fertilizer is also applied in spring.

There is scarcely any of the border that does not contribute to the spring display. Tulips, well suited to the heavy soil, provide the highlights and remain tidy when the leaves of narcissi have become disagreeably lank. Sometimes they are in permanent groups growing through herbaceous plants like *Hemerocallis lilioasphodelus* (*H. flava*) and Japanese anemones, whose leaves take over as the bulbs die back; sometimes they nestle below shrubs such as hydrangeas and summer-flowering tamarisk which grow out to cover the dying leaves as spring moves into summer. Less permanent sorts are planted out each year and replaced by annuals like *Cosmos* 'Sensation' or tender perennials such as cannas.

Christopher Lloyd admires the way low-growing spring flowers are woven through the

Purple Border at Sissinghurst, even to the back, their places filled by bulkier perennials as the season advances. At Dixter, pulmonaria and *Omphalodes cappadocica*, not very light-demanding and happy to grow at the back of the border, show their sky-blue flowers before herbaceous plants have grown. Elsewhere in the border snowdrops are followed by violets, primroses and grape hyacinths and vivid royal blue irises push through the startling yellow-green leaves of *Geranium* 'Ann Folkard'.

For summer, the delphiniums and dahlias favoured in the original planting were too time-consuming to keep in good order and have been replaced largely by summer-flowering shrubs like buddleja and summer tamarisk. These can be treated like herbaceous plants, cutting them back hard every year. Numerous roses are worked into the border in this way, the other plants hiding their ugly ankles. The large, scarlet-flowered shrub rose 'Florence Mary Morse' is valued for giving red in the border over a long period, something not many herbaceous perennials can do. Christopher Lloyd prefers roses that have a double season, like *R. moyesii*, and *setipoda* that are pretty in flower and, later, in fruit.

Perennials such as phloxes are used for big patches of exciting colour in the middle of summer but foliage plants like the cardoons have played an increasing role, handsome in leaf and with the bonus of flowers later. Plants that will integrate a border by growing into their neighbours, like geraniums and violas, are favoured and Christopher Lloyd has no time for gardeners, especially professional gardeners, who go to great lengths to keep their plants separate, going around each plant with a hoe so that there is a distinct gap between them. Magenta 'Russell Prichard' and 'Ann Folkard', the same colour but with a glistening black eye and yellow-green spring foliage, are among the longest-flowering and best of the rambling geraniums.

Plants that flower early and then die back, such as alstroemerias and *Allium neapolitanum*, can have their tops quickly removed and be planted over with annuals. In other cases, continuity of flower is achieved by dead-heading, as for *Salvia* × *superba* and heleniums. Plants that tend to get scruffy by late summer, such as *Hemerocallis lilioasphodelus*, *Geranium* × *magnificum* and *Alchemilla mollis*, are cut off ruthlessly ; their new foliage soon appears and remains fresh for the rest of the season.

Climbers help provide the vertical accents, either trained up columns or scrambling over shrubs, and take up little space. Clematis are much in evidence and can be fitted in 'as much as you can be bothered': 'Jackmanii Alba' scrambles into the silver willow and 'Jackmanii Superba' into golden elder whose acid yellow-green combines with pink monarda and phlox to make a typically sharp mix of colours. A rare plant of what is

believed to be the true *Lonicera* × *americana*, ousted from the British nursery trade by the more easily rooted *L.* × *italica*, is grown up a pole in this way. Lilacs, considered boring but kept because of sentimental attachment, also add height and provide support for swags of climbers such as *Parthenocissus inserta*, briefly glorious in autumnal red; sometimes a large old limb needs to be pruned out of the lilacs to keep them vigorous and full of blossom.

The colour of foliage is as important as that of the flowers: dark *Berberis* × *ottawensis* 'Superba' provides a convenient peg for the everlasting pea *Lathyrus grandiflorus* and creates a harmony with the two-toned magenta-pink and maroon flowers of the pea and nearby *Digitalis* 'Sutton's Apricot'. *Clematis recta* 'Purpurea' adds an element of surprise, particularly in spring when its leaves

Underplanting of spring bulbs is an important feature of the Dixter border. Most of the tulips are yellow, orange or creamy white, colours that sparkle in a sea of self-sown forget-me-nots which also provide an effective foil for the bold leaves of Allium christophii. *In the picture, extravagantly striped 'Absalom', an ancient Rembrandt variety, is seen behind a subtly shaded and elegantly shaped cultivar brought back from a Mediterranean holiday, witness to Christopher Lloyd's eye for a good plant. The foliage of the smoke bush will not envelop the tulips beneath until they have sufficient reserves for next spring's flowers.*

are most richly coloured (staking is essential and it needs brushwood at least 1.8m/6ft tall inserted at the last possible moment before the plant flops). *Ligustrum* 'Vicaryi' is chosen for its soft yellow feathery foliage, altogether less brash than the common golden privet.

Another floppy favourite that more than justifies the task of staking is *Aster sedifolius* (formerly *A. acris*). Nasturtium is planted to trail through it and stud the mauve daisy flowers with sparkling red, most effectively when it is not stunted by a dry summer.

There are too many gardeners who eliminate from their borders any plant that requires a little extra effort. The result is usually dull. This is not so in the borders at Dixter where many of the highlights are provided by plants that have to be staked or replanted each year or raised annually from seed. Even invasive plants are not excluded if they earn their keep. Blue lyme grass, *Leymus arenarius* (syn. *Elymus arenarius*) was once confined to an old tin tub formerly used for bathing dogs but it sulked for years; now given the run of the border, its questing roots have to be traced back to its allotted space each spring. It takes a brave gardener to let loose *Houttuynia cordata* 'Chameleon' but at Dixter it is kept in check by being baked in full sun at the front of the border, quite the reverse of the moist shady conditions it prefers; such unconventional treatment gives the variegated leaves an even more spectacular scarlet flush and proves Christopher Lloyd's belief that 'sometimes you shouldn't follow the maxim of doing what's ecologically correct'.

The occasional tall plant, such as a 1.8m/ 6ft high clump of brilliant sky-blue *Salvia uliginosa*, lurches to the front of the border. The effect is slightly unnerving: like a clown on stilts, one wonders how it manages to stay upright and worries about its imminent collapse. But this is a deliberate policy to show to their best advantage plants of thin texture or tall and graceful habit, which could not be appreciated if buried deep in the border.

Verbena bonariensis is used in this way and grasses such as *Arundo donax* and variegated miscanthus, although little used here, are treated thus in Great Dixter's Barn Garden and elsewhere where they make wonderful fountains of foliage. Gardens where such beauty is submerged are dismissed with scorn.

When asked which borders in other gardens he dislikes or which he would have planted differently, Christopher Lloyd's answer is 'Practically all of them!' This is not a criticism of all the rest but is an assertion of his own strong and individual preferences, seen in every part of his garden. At Dixter the impedimenta of border-making, the perceived rules accumulated over 200 years, have been defiantly jettisoned. Never mind what anyone else says or thinks about how a border should be composed, here Christopher Lloyd has worked from his own first principles to create the sort of effect he wants.

Of course, not everyone will share his taste. There will be some plant associations that are not a total success; but for every failure there are several groupings of two or three plants in which the beauty of the combination far transcends the attraction of the individual. And there will always be another year in which the planting can be changed and new associations attempted. It is this continual reworking, ever striving for a better effect, that keeps Great Dixter, and every other first-rate plantsman's garden, constantly fresh and alive.

OPPOSITE *Sumptuous clematis and 'Royal Purple' smoke bush add weight and richness to the back of the border. Domed golden heads of* Senecio doria *provide contrast, their colour echoed in spikes of* Verbascum nigrum. *In front, tiny florets of* Hydrangea arborescens *'Annabelle' massed into gigantic heads glow cream and palest green against the sun.*

RIGHT *Typically bold use of contrasting colour and form sets airy heads of* Verbena bonariensis *and slender spikes of* Buddleja *'Lochinch' against weightier 'Royal Purple' smoke bush and glowing* Helenium *'Moerheim Beauty'; the powerful verticals of kniphofia, tipped with fiery orange, promise to add to the drama.*

INFORMAL EARLY SUMMER BORDERS

The Manor House, Heslington

George Smith's double borders in the Old Shrub Rose Garden at the Manor House, Heslington, are utterly informal: the planting is irregular, the two sides are far from identical and the path between them has a distinct wiggle. They are much the earliest in *Best Borders*, planned so that the peak of their display begins with the tulips and ends as the old roses fade. George Smith is everything one would expect of a flower arranger of international repute: artistic, witty, charming and flamboyant. He has a canny regard for the qualities of texture, colour and form of all the plants he uses, both in his displays and in his garden. All these characteristics are apparent in his arrangements, though it must be owned that these have an extravagance and perfection that make them seem beyond the world of the man or woman in the street and

more fitted to palaces or cathedrals: at one of his demonstrations, a lady in the audience asked if he did anything for bungalows.

George Smith came to the Manor House at Heslington near York in 1968 and took over a garden that Nina, Lady Deramore, herself a noted flower arranger, had gardened since 1947. The house had been home farm to Heslington Hall and the part of the garden where Lady Deramore had grown her old roses, called then the Paddock and now the Old Shrub Rose Garden, had been the stackyard of the farm. Topsoil was brought and tipped by tenant farmers on top of the layer of river pebbles that formed the surface of the yard; this pan of hardcore prevented water draining away, giving the impression that the subsoil was heavy clay, though in fact it is a light sandy loam.

While many borders will continue in bloom for some time after the roses have finished, the double borders in the Old Shrub Rose Garden at Heslington Manor are unusual in their early flowering. At the height of the display, the large, rounded flowers of peonies and oriental poppy 'Sultana' are echoed in the distance by vibrant rose 'Cerise Bouquet'. Spikes of delphiniums

and lupins provide contrast while smaller-flowered cranesbills, Geranium × magnificum *in the foreground, magenta* G. psilostemon *and varieties of* G. pratense, *provide a foil for the larger blooms. The strikingly architectural cardoon, furnished with immaculate grey-green leaves from top to toe, provides a bold contrast of form.*

On this unpropitious site, Lady Deramore assembled roses from Sunningdale Nurseries, helped by advice, visits and frequent correspondence from Graham Thomas. The roses were planted through turf but because the ground had not been properly prepared most of them died.

With the help of the original gardener, Norman Puckering, George Smith started to replant the roses in the mid 1970s, hacking out large holes about a metre/yard square and a metre/yard deep, filled with compost and topsoil from the vegetable garden. 'There is really no decent soil,' he says; 'every time we plant something, we *have* to improve the soil; penny for the plant, pound for the hole.'

The roses there date from this period and their pink, purple and magenta colouring has provided the theme for what has become a splendid mixed double border, at the peak of its display in early summer. Hidden among encircling shrubs and ornamental trees, the borders are a delightful surprise to the visitor, separate and secret, seen only as the most tantalizing glimpses from without.

This air of secrecy is provided by an archway formed by *Prunus × subhirtella* 'Autumnalis' at one end, echoed by another at the opposite end provided by *Rosa* 'Carmenetta'; weeping across the path, this marks a transition into a shadier, almost woodland, area surrounding a pool, a part of the garden with a quite different character.

Other shrubs provide the billowing walls of this outdoor room, for instance the snowball bush, *Viburnum opulus* 'Roseum', and *Rosa xanthina* 'Canary Bird'; yellow here is felt to be wrong, though George Smith is loath to get rid of it. However, he is usually ruthless about colour: a large bush of rose 'Frühlingsgold' has recently been sacrificed. Between some of the shrubs, a squint across lower herbaceous plants entices the visitor with a view of the exuberant *floraison* within.

Though the path between the borders is more or less straight, the effect is utterly informal. A large bush of *Rosa nutkana* 'Plena' (sometimes called *R. californica* 'Plena') growing 3.6m/12ft high at the front of the border adds to the informality; some say that every double border should have its one glorious eccentricity but George Smith wishes he could move this back a metre/yard or two. The rose tends to divide the border one-third of the way down, seeming like a large obstacle; however, it does have the advantage of creating a separate niche where some choice planting can be shown off to advantage.

An important early task was the planting

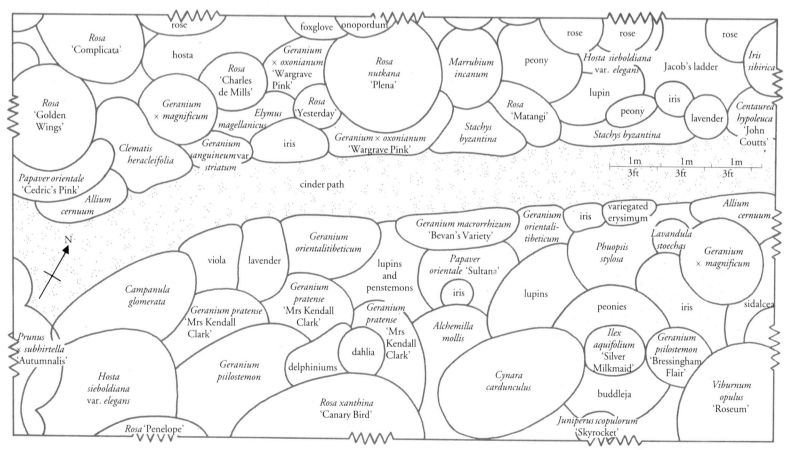

of a hedge between the rose garden and the vegetable garden, both for shelter and to provide a background to the planting. Beech was chosen, partly because George Smith was too impatient for yew, also because yew is a greedy feeder, competing more aggressively with nearby plants. Besides, beech seemed right for the surroundings, its winter russet complementing the hand-made brick walls, its pale spring green equally attractive. The coarser texture and slightly rustic, informal character seem much more appropriate too than the severe smartness and formality of yew.

The roses were chosen largely from the descriptions in Graham Thomas's *The Old Shrub Roses* but as they matured they became bare and leggy. Training onto broad crinolines of hazel hoops, as at Sissinghurst, was not an option here: there simply was not the room. Some custom-made iron supports provided

by a local blacksmith have been used to train the roses as splendid columns and prevent them from smothering the herbaceous plants. Even with these, and given expert pruning, it will still be difficult to keep the roses furnished to the base. George Smith realized that he would need plants to enhance the roses and decently clothe their ugly ankles. These he collected bit by bit, with colouring to match, many of them brought from other parts of the garden.

It does not worry George Smith that the roses are not repeat flowering: tucked away in a corner of the garden, a lack of flower here can be excused; by the time these borders have completed their main display there is plenty of other interest elsewhere. The main consideration is that as many as possible of the plants that accompany the roses should flower at the same time.

OPPOSITE *The fullness of the borders disguises the fact that they are not of equal depth. Roses trained to narrowly upright supports on the northwest side of the path (shown top) conceal the hedge so that the viewer is not aware of the closeness of the boundary. The numerous varieties of cranesbill would not be sufficient foil for rose and peonies without some variation in foliage texture, provided here by hostas, blue-green* Elymus magellanicus *and bearded irises. A large* Rosa nutkana *'Plena' (*R. californica *'Plena') towards the front of the border creates a separate bay in the planting and blocks the distant view, encouraging the viewer to explore further down the path.*

BELOW *As the peonies and* Rosa nutkana *'Plena' pass out of flower, a few stalwarts continue the display into midsummer: richly coloured* Campanula lactiflora, *rose 'Matangi' and lupins.*

Because the planting includes irises, peonies, poppies and lupins, there is no June gap at all. The display continues in full glory from late spring to early summer; by mid-summer the main display is over. However, this does not mean that there is no interest at other times. Tulips can always be fitted in for an earlier display: there are pockets of double pink peony-flowered 'Angélique', the Viridiflora 'Groenland', 'Queen of Night' and 'Blue Parrot' and extravagantly ruffled 'Black Parrot'. In spite of the rock-hard soil, the tulips will grow in spaces filled later by perennials or pierce through carpets of plants such as *Phuopsis stylosa*, the pink bedstraw.

For later in the season, there are dahlias such as 'Arabian Night' and plenty of penstemons, especially lilac-coloured 'Alice Hindley' and 'Stapleford Gem', often misnamed 'Sour Grapes', a fascinating mix of turquoise-blue and lilac-purple, sometimes shading almost to pearly white.

The principles for combining plants in

LEFT *An underplanting of tulips, including sultry 'Queen of Night', provides late spring colour before the borders' main season, the deep magenta of* Geranium macrorrhizum *'Bevan's Variety' chiming in unison.*

OPPOSITE, LEFT *Rose 'Cerise Bouquet', seen here with* Berberis thunbergii f. atropurpurea *and* Knautia macedonica, *can instantly be recognized by the many bracts on its flower stems, inherited from* Rosa multibracteata.

OPPOSITE, RIGHT *Though it is intolerant of drought,* Geranium clarkei *'Kashmir Purple', seen here with* Centaurea hypoleuca *'John Coutts' and catmint, has a poise and charm altogether lacking from some of its more showy kin.*

OVERLEAF *The enchantment of the Heslington borders at their most irresistible, as the rays of the morning sun creep above the encircling shrubs.*

George Smith's garden are the same as those for his displays: 'my border is a flower arranger's idea of a *living* flower arrangement, so the colours and the textures that I would put together in a vase of flowers I apply to the border.' Apart from the writings of Graham Thomas, two artist-gardeners have been his main influences. Beth Chatto's books and inspirational displays at the Chelsea Flower Show have influenced a generation of gardeners. George Smith says Mrs Chatto 'is a flower arranger, although she might not admit to it; she started as a flower arranger and gardens with an eye to colour but also with an eye to form and texture of foliage'. These principles are applied equally to the borders at Heslington.

The other important influence is John Treasure, whom George Smith describes as an artist: 'his use of colour combination is so striking. I remember seeing clumps of verbena in a dry hot area that he'd put with clumps of *Sedum telephium* subsp. *maximum* 'Atropurpureum'. That was the first time I'd really become aware of the possibility of using coloured foliage.' Through John Treasure's example, many plants have been added solely for the effect of their foliage.

The first important plant added to the roses was *Papaver orientale* 'Sultana' from Beth Chatto. In a curious way, this has proved to be the linchpin of the design, affecting the choice of many of the plants subsequently added. George Smith considers it a difficult colour, a hot cerise pink. He felt that cranesbills would combine agreeably with it, as well as with the roses. As in a flower arrangement, larger flowers benefit from the addition of smaller ones; repetition of the leaf and flower shapes of quite a collection of cranesbills in various colours and statures gave unity to the design, the geraniums flowering at the same time as roses and poppies. George Smith was not alone in his enthusiasm for cranesbills:

many gardeners followed Graham Thomas's advocacy of them, making them by the mid 1980s by far the most popular hardy herbaceous genus, according to a Hardy Plant Society survey.

Introduction of the 'quite hectic' *Geranium psilostemon*, magenta with a black eye, demanded yet another addition to the scheme. John Treasure's example of the use of grey, silver and glaucous-blue foliage was followed, an effective means of toning down the fiercest pinks and magentas as well as providing an agreeable foil for other pinks, blues and purples. As already noted, Graham Thomas has pointed out that the reason Gertrude Jekyll disliked magenta and avoided using it could well be that few grey-leaved plants were available in the early twentieth century; it was not until they were championed by gardeners such as Mrs Desmond Underwood that their usefulness became widely accepted and understood.

Some would argue that by the 1970s grey foliage was used with gentle colours almost to excess, creating schemes of suffocating softness and all but banishing more daring designs contrived for richness or dazzle. So a spectacular poppy had been added to the roses; this had demanded hardy geraniums in considerable variety, worked in throughout the border to unify the design; then grey foliage was added to temper the hottest colours and provide a restful foil for all the flowers. Thus the planting evolved logically and step by step until there was enough variety of texture, colour and form, but, as with a flower arrangement, never so much that the composition became an incoherent jumble.

The milk-white blue-veined flowers of *Geranium pratense* 'Mrs Kendall Clark' pop up throughout the border wherever they sow themselves and combine happily with every other colour and leaf texture, even chiming in harmony with the difficult *G. psilostemon*. *G. × magnificum* is also repeated through the borders; I find the unrelieved solidity of its lavender-blue flowers overdominant and unbalancing, especially one very large group at the front of the border that seems to upstage every other plant. *G. clarkei* 'Kashmir Purple' was a charmer for some years, its flowers more modestly spaced and borne over many weeks, but sadly it proved unequal to the hot dry summer of 1992.

The greys and blues include 'good old lamb's ears' and 'that grass that keeps changing its name', *Elymus magellanicus* (syns. *Agropyron magellanicum, A. pubiflorum*). George Smith agonizes over his cardoons, wondering whether he should let them flower and suffer their subsequent shabbiness or cut them down for a new crop of handsome leaves. Onopordums have their moment of glory at the same time as the roses and add a theatrical touch: 'you need exclamation marks and punctuation. It's a bit like writing prose, in a way: you have the odd purple passages.'

Glaucous hostas are important for both foliage colour and form, including the magnificent *H. sieboldiana* var. *elegans*, a particular favourite, then *H. fortunei* var. *hyacinthina* with a narrow silvery-white edge to a glaucous leaf, the Tardiana Group and a newcomer from Wisley, 'Blue Lotus Leaf'. However, variegation is avoided for two reasons: it tends to be cream and green (few if any hostas have glaucous foliage variegated with white); cream has yellow in it and so is banned from this part of the garden; secondly, variegation here would seem rather fussy.

Other important contributors to the borders' main display are delphiniums and lupins, valuable for their vertical accents, the assorted cultivars giving slightly staggered flowering times. Peonies echo the roses in flower colour, scale and form, even to some extent in their foliage, bringing these qualities to the front ranks of the border; the foliage is particularly attractive when young, fresh and glossy, flushed with red and full of promise. Bearded irises are also essential, their sword-like foliage providing strong contrasts and their gentle and varied fragrances of primroses, marzipan or violets adding another dimension.

Taller campanulas are used to hide the bare bases of shrub roses and are also useful for cutting; George Smith prefers all his plants to last in water and there are not many throughout the garden that cannot also be used in arrangements. *C. latiloba* 'Highcliffe Variety' is particularly good; new *C.* 'Elizabeth' with large, speckled dusky pink trumpets promises to be another prized performer.

The borders have no rigid gradation of heights. Some tall plants approach the front while other short ones are woven through the border towards the back. George Smith apologizes for this lamentable lack of discipline, though he need have no qualms: irregularity in height is in keeping with the borders' informality and is part of their charm.

To those not familiar with the borders' history as a garden of old roses, the roses seem not to predominate; the borders seem to be particularly well-balanced, with roses playing no more important a role than several of the other plants, cranesbills, irises or peonies.

The borders are at their most magical in the soft and misty light of early morning; looking away from the house, the rays of the sun illuminate the tallest flowers, throwing them into startling prominence against a shadowy background. Viewed from the opposite direction, the effect is altogether different, gentler and with less contrast of light and dark. The endless possibilities for a different play of sun and shade as the day or the season progresses and the change in light quality depending on whether the border is seen from one direction or the other are the great advantages of double borders over those that are one-sided or hidden from the sun. Such characteristics are seldom given the consideration they deserve and should be much more widely exploited: flowers lit from behind, particularly those in warm colours, can glow with astonishing intensity, and fine-textured plants such as grasses can shimmer and sparkle magically.

George Smith has built up the planting of these borders as he would create a flower arrangement: starting with the roses, other plants have been added to provide contrasting texture and complementary colour one by one until there are just enough. The perils of plantsmanship – adding too many different and discordant varieties – have been avoided. This approach is different from that of garden designers, who so often decide on the mix of colours and textures, then set down the whole design at once, a method that frequently gives a less successful result than a more evolutionary process. At Heslington, George Smith has shown how successful the flower arranger's approach can be, and it is a method that will suit many other gardeners, whether they be amateur or professional.

LEFT *Framed in an arch of rose 'Carmenetta', irises, alliums,* Hosta fortunei var. hyacinthina *and* Geranium clarkei *'Kashmir Purple' provide gentle colour and an interplay of textures in the early days of the borders' display.*

RIGHT *The cardoon at its moment of greatest glory: with such striking architectural form there is little need of colour. The fat and burgeoning buds of peonies promise future splendours beside the bristling and agreeably frantic 'Silver Milkmaid' holly; behind, 'Canary Bird' is the harbinger of the roses.*

BORDERS IN A SMALL TOWN GARDEN

Oxford

Anne Dexter's garden shows the best possible use of limited space. She admits that her passion for plants is out of all proportion to the size of her garden which is the width (6m/20ft) of her town house in Oxford and 23m/75ft long. There can be no doubt that her solutions to the limitations of space have brought rich rewards and a garden tightly packed with treasures. Her garden succeeds not merely because it contains the choicest plants in abundant variety: the basic structure is good, with plenty of evergreens to ensure substance throughout the year; foliage colour is used to the full; climbers give both height and late colour. Perhaps the most useful lesson it demonstrates is that if you cannot garden outwards, garden upwards. Even in the tiniest of plots, the sky is the limit.

Anne Dexter calls her garden 'a long thin rat run'. It is mysteriously tunnel-like and reminiscent of the rabbit hole in Alice's Wonderland, with a tantalizing false door at its furthest end. Only the uppermost shrubs and clematis reach high enough to benefit from full sun; except in summer, the sun seldom penetrates to the shorter plants. Though the tall clematis-swagged shrubs give exactly the effect she wanted, she is no longer so fond of teetering on top of step ladders to train and prune them as she was thirty-odd years ago. The furthest section of the garden is shaded by several small trees; here only the most tolerant of species will thrive in the sepulchral gloom. The problems of the small garden, with much shade and tall, narrow borders almost assuming the character of a

Anne Dexter's garden is remarkable not only for its skilful use of a small site but for the immense variety of plants it contains. Its borders probably contain more varieties than any other featured in this book. The path is slightly kinked so that the viewer is never sure what lies beyond and is tempted to explore. Miller's grape, Vitis vinifera 'Incana' is neatly trained above the french windows while, beneath,

the blue satin flowers of Convolvulus sabatius (syn. C. mauritanicus) flower all summer long, fully justifying the winter protection they need. Dark-stemmed Dahlia 'Honor Francis' stands sentinel beside the path. The idea of turning the perimeter border of a tiny garden into a tall and broad hedge of varied flowers and foliage like this could be copied by many gardeners.

117

richly diverse flowering hedge, are familiar to many gardeners but have rarely been so successfully tackled.

Anne Dexter comes from a family of keen gardeners: her mother and grandmother both gardened, though she is the first of her family who has done the work herself rather than delegating it. Her daughter shares her obsession with plants and accompanies her on well-planned tours of the finest nurseries. Each of these circuits results in a haul of precious new plants for which there seems to be not the least room for planting, yet somehow they all get squeezed in.

Before she moved to her present house in 1957, Anne Dexter had been used to a much larger garden. She realized from the first that every scrap of ground would be needed for plants and there would be no space for lawn. The fences around the garden were heightened with trellis, making the planting seem steeply banked on both sides. The path was excavated, putting in steps and edging alongside the path with dwarf walls, thus making the height of the sides seem even more dramatic. This herculean task (all the excavated soil had to be removed through the house) was the last major project in the making of the garden, though the planting has been constantly amended ever since. Many of the slow-growing shrubs and small trees put in to form the bones of the planting have

LEFT *In a canyon of towering borders walled with the finest plants, small trees crest the ridge, their leaf colour as attractive as flowers but longer lasting. Elaeagnus 'Quicksilver', Dickson's golden elm, purple plum and* Robinia pseudoacacia *'Frisia' all give bold display from spring to autumn as well as providing support for clematis. Tall shrubs such as* Clerodendrum trichotomum, *photinia and* Hydrangea aspera subsp. sargentiana *make up the next layer of planting, with upright perennials such as* Euphorbia sikkimensis *and* Lobelia × gerardii *'Vedrariensis' beneath. Ivies and euonymus furnish the dwarf walls flanking the path, with* Sedum telephium subsp. ruprechtii, *diascias and osteospermum in the foreground.*

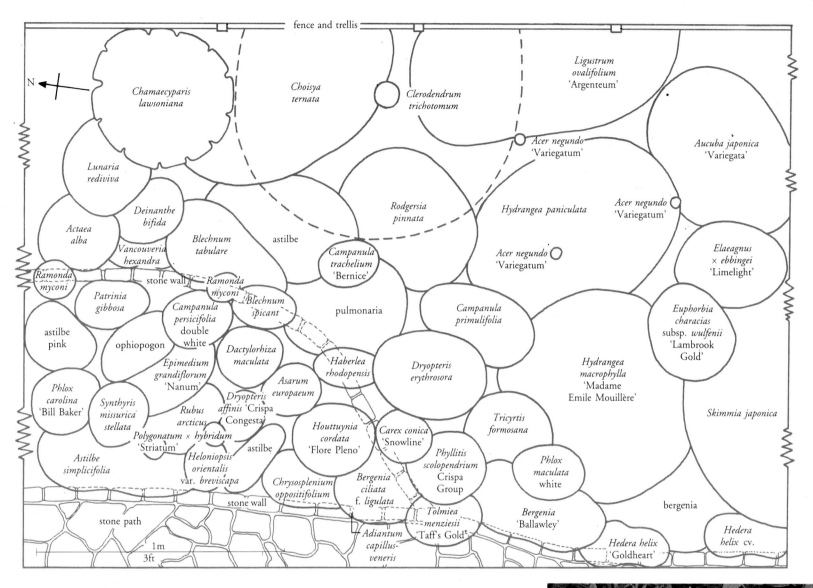

fence and trellis

N

Chamaecyparis lawsoniana

Choisya ternata

Clerodendrum trichotomum

Ligustrum ovalifolium 'Argenteum'

Acer negundo 'Variegatum'

Aucuba japonica 'Variegata'

Lunaria rediviva

Rodgersia pinnata

Hydrangea paniculata

Acer negundo 'Variegatum'

Deinanthe bifida

astilbe

Acer negundo 'Variegatum'

Elaeagnus × ebbingei 'Limelight'

Actaea alba

Blechnum tabulare

Campanula trachelium 'Bernice'

Vancouveria hexandra

Ramonda myconi

stone wall

Ramonda myconi

Blechnum spicant

pulmonaria

Campanula primulifolia

Euphorbia characias subsp. *wulfenii* 'Lambrook Gold'

Patrinia gibbosa

Campanula persicifolia double white

astilbe pink

ophiopogon

Dactylorhiza maculata

Haberlea rhodopensis

Dryopteris erythrosora

Hydrangea macrophylla 'Madame Emile Mouillère'

Epimedium grandiflorum 'Nanum'

Asarum europaeum

Phlox carolina 'Bill Baker'

Synthyris missurica stellata

Dryopteris affinis 'Crispa Congesta'

Skimmia japonica

Rubus arcticus

Polygonatum × hybridum 'Striatum'

Houttuynia cordata 'Flore Pleno'

Carex conica 'Snowline'

Tricyrtis formosana

Heloniopsis orientalis var. *breviscapa*

astilbe

Phyllitis scolopendrium Crispa Group

Phlox maculata white

Astilbe simplicifolia

Chrysosplenium oppositifolium

Bergenia ciliata f. *ligulata*

Bergenia 'Ballawley'

bergenia

stone wall

Tolmiea menziesii 'Taff's Gold'

stone path

Adiantum capillus-veneris

Hedera helix 'Goldheart'

Hedera helix cv.

1m

3ft

ABOVE AND RIGHT *At the shady end of the garden, a dwarf wall separates a plateau at the front of the border housing select woodlanders and ferns. The wall allows gesneriads* Ramonda myconi *and* Haberlea rhodopensis *to grow on their sides, shedding the winter wet they detest. The background consists of variegated aucuba, elaeagnus, ligustrum and acer to lighten the darkness, aided by white flowers of* Hydrangea macrophylla *'Madame Emile Mouillère' and* H. paniculata. *Three plants of variegated box elder (*Acer negundo *'Variegatum') have only a fraction of the space they would usually need and all the plants have to jostle for space. Though there are berries (doll's eyes,* Actaea alba*), seed pods (*Lunaria rediviva*) and flowers in plenty,*

notably unusual double bellflowers, it is foliage that dominates. There is the fullest use of it in a wide range of shapes and sizes: here are grassy ophiopogon, carex and heloniopsis, ferns such as dryopteris, blechnum, adiantum and phyllitis and big bold leaves of rodgersia and bergenia. B. 'Ballawley' (on the right of the photograph), for which nurseries often substitute inferior seedlings with less magnificent foliage, is among the most splendid of these, seen here with the immaculately goffered leaves of Phyllitis scolopendrium *Crispa Group, silver-splashed pulmonaria,* Dryopteris erythrosora *and* Carex conica *'Snowline'. The fronds of the dryopteris are flushed rosy red as they unfurl.*

since proved to be anything but miniature: Anne Dexter now has to garden upwards because there is simply nowhere else to go.

The key plants in the garden are chosen to give year-round interest, with a considerable proportion of evergreens and abundant gold and maroon foliage distributed throughout. Even during the winter the conifers provide boldly structural shapes in green and gold, with ivies clothing the fences and other evergreens such as box, elaeagnus, euonymus and bold-leaved fatsia contributing to the effect. Among the purples, *Berberis thunbergii* 'Somerset', *B. t.* 'Golden Ring' and *Cotinus coggygria* 'Foliis Purpureis' are especially good value. Anne Dexter is uncertain whether she likes *Prunus cerasifera* 'Pissardii', planted to give accustomed shade after the untimely collapse of all the trees at the end of the garden when they blew down in a single night: its solid outline and deep colour cannot give quite the light and subtle effect that was intended, though it does continue the theme of purple around the end of the garden. It is pruned and thinned every year but still tends to lumpishness by the end of the summer, the intensity of its tone almost too contrasting against golden *Robinia pseudoacacia* 'Frisia' nearby.

Ulmus 'Dicksonii' echoes the gold of 'Frisia' on the opposite side of the path while among gold-leaved shrubs, *Lonicera nitida* 'Baggesen's Gold' plays an important supporting role, carrying a large and floriferous plant of *Clematis × durandii*. Yellow is also provided by the dogwoods *Cornus alba* 'Spaethii' and *C. a.* 'Aurea' which have the bonus of red stems in winter. Some silvers are also prominent, particularly *Elaeagnus* 'Quicksilver', used to support a number of clematis, notably the Viticella hybrid 'Elvan'.

Variegation helps lighten up a dark corner but is allowed only in moderation, preferably with leaves neatly edged in silver or gold, not untidily blotched: silver privet is a favourite

but *Berberis thunbergii* 'Rose Glow', with purple leaves splashed cream and pink, is despised and not grown, even though Anne Dexter collects other *Berberis thunbergii*. Every flower colour is found except orange: even the most exquisitely formed bloom of this hue would not be permitted; however, rich apricot *Caiophora lateritia* is deemed acceptable, in spite of its stinging hairs.

The abundance of colourful foliage through the winter causes the transition to spring to be less noticeable than usual, particularly since there are few spring flowers and almost no bulbs. These attract the constant attention of herds of browsing slugs and most would not thrive in the darkest depths of this canyon of plants. Repeated attempts to grow crocuses have failed and tulips are felt to be too transient to be worth the cost and effort of growing them.

The borders contain relatively few medium-sized herbaceous plants, their place being taken by small shrubs that are more able to support the many clematis varieties. Those there are seldom need to be divided: the close planting prevents even potentially invasive varieties such as the pink and white forms of rosebay willowherb from spreading rampageously. Herbaceous clematis such as *C. integrifolia* var. *albiflora*, *C. i.* 'Pastel Blue' and *C. i.* 'Rosea' have been a disappointment and produced few flowers. Anne Dexter has little time for plants such as these, however choice and unusual, that do not pull their weight.

The fronts of the borders are occupied predominantly by alpines in the sunnier parts of the garden near the house and elsewhere by woodlanders, plants of modest size that will grow equally well in the border or the dwarf walls flanking the paths. Among the alpines, campanula, erodium and origanum are clearly among the favourite genera. Though Anne Dexter is not particularly fond of the varieties of *Euonymus fortunei* used alongside the path, she feels they are useful

and tolerant of being flopped upon throughout the summer by the likes of long-flowering *Geranium wallichianum* 'Buxton's Variety'. The cranesbills are well represented throughout the garden, not only smaller sorts but taller varieties such as the extravagantly striped *G. pratense* 'Striatum' and scramblers such as magenta *G. procurrens*.

Ivy-covered poles provide bold vertical accents at intervals on both sides of the garden. Created from 3.6m/12ft saplings with side branches stubbed back to 15cm/6in, these have been wrapped with brown plastic mesh, to train not only the ivies but yet more clematis which climb up them through the summer months. Any shrubs not thoroughly woven through with clematis are draped with some choice scrambler, a species of lathyrus such as yellow-green flowered *L. chloranthus* or golden *Dicentra macrocapnos*. Another favourite climber is *Vitis vinifera* 'Incana', miller's burgundy grape, carefully trained and pruned above the french window where Anne Dexter can enjoy watching the birds eating the two hundred or so small bunches of grapes it produces each year.

Anne Dexter treasures her many plants not just for their individual beauty but for their associations with gardening friends. Percy Picton, a renowned plantsman and nurseryman who had worked for William Robinson and knew Ellen Ann Willmott, provided numerous choice varieties in the garden's early days. Arthur Branch, commemorated by a marvellous variety of *Sedum telephium* with maroon leaves and glowing red stems, always had some rare and exquisite alpines. Both are remembered sitting in their sheds taking cuttings surrounded by the reek of tarry smoke from the stove. Some of their plants still survive in her garden as a memorial to them, though Anne Dexter admits that 'most have gone the way of all things; they don't last for ever.'

Plant hunting forays follow well-tried routes to nurseries. Anne Dexter often sees

plants she feels she cannot live without, though she tries to resist the temptation of tender plants which might be fine in Zone 9 but not in her frost-pocket garden in Zone 8. She is philosophical about the occasional losses, in a way essential to her style of gardening if new plants are to be accommodated. At least plants which do not survive in the long term will give pleasure for a season or two before they succumb, engulfed by their neighbours which crowd in on them from every side. Finding plants to put in the middle or back ranks of the border to replace any losses can be difficult: the newcomers need to be 1.2–1.5m/4–5ft to begin with if they are to hold their own.

Plants at the furthest end of the garden have to be true shade-lovers to thrive. These include species like *Fritillaria pyrenaica* which make the most of spring sunshine before they are adumbrated by the canopy of deciduous

trees. This fritillary is a particular favourite, brought from Anne Dexter's mother's garden outside Kirkby Lonsdale in Cumbria. Its rich purple-brown nodding flowers can be turned up to reveal an intriguing centre of purple with glistening yellow nectaries at the base. Other plants that flourish in the gloom include anemonopsis, doll's eyes (*Actaea alba*), tricyrtis, trilliums and a range of the finest epimediums.

At the front of the border is a relatively large flat semicircular area backed by another dwarf wall; here is a collection of the choicest but tiniest plants including a number of ferns, most of them from Reginald Kaye's nursery at Carnforth in Lancashire. Mr Kaye was renowned for the wide variety of ferns he grew; the very best clones, with fantastically ruffled blades and criss-crossed or congested fronds, are slow to increase and plants were seldom offered for sale. It is evidence of Anne

TOP LEFT *Cranesbills are a favourite genus here, most of them hardy in Zone 8, not demanding full sun and easy to grow.* G. pratense *'Striatum', its white flowers splashed and streaked with lavender-blue, is one of the most freakish. Others such as* G. procurrens *and its hybrids 'Ann Folkard' and 'Salome' are scramblers, a class of plant that Anne Dexter uses widely to drape her shrubs.*

BOTTOM LEFT Geranium kishtvariense, *plantsman Roy Lancaster's recent introduction from Kashmir, is valued for its rich pink flowers.*

ABOVE *Diascias have become immensely popular among gardeners in Zone 8; many species are almost completely hardy there while others are tantalizingly not quite so.* D. vigilis *has proved worth the risk. Here its slender stems are held up by sprigs of bamboo. Soft salmon-pink flowers harmonize with the native* Papaver dubium *and contrast agreeably with campanulas, nigella and* Eryngium bourgatii *'Oxford Blue'.*

Dexter's very considerable charm and powers of persuasion that so many of them found their way to Oxford. The dwarf wall at the back of this area houses ramondas and haberleas, unusual among the Gesneriaceae in being relatively hardy, though the flowers have the look of exotic tenderness of their cousins the African violets. Happier growing on their side than flat on the ground where winter wet can rot their crowns, the lavender or white flowers of these gesneriads are a delight.

Though the many clematis varieties provide colour through late summer and autumn, most of the shorter plants flower in spring or early summer. Some annuals are added at the front of the borders to give continuity of display into the latter half of the year. Dwarf green tobacco flowers and blue lobelias are used to fill spaces along with white *Cosmos bipinnatus* 'Sonata', a particular favourite. At 60cm/2ft, this is more useful than the taller sorts for the small garden, without being so dwarf that its natural grace is lost. One larger plant bedded out each year

OPPOSITE *There is scarcely a shrub or small tree in the garden that is not host to a scrambling plant of some sort. Geraniums, dicentras and perennial peas all play a role but by far the most widely used are varieties of clematis. The Viticella cultivars are among the most useful, easy to prune and finding their way to the top of the planting each year with the minimum of help. 'Alba Luxurians' and* C. viticella *'Purpurea Plena Elegans'* (TOP LEFT) *combine well with purple plum and variegated box elder, while 'Minuet'* (BOTTOM LEFT) *copes happily with shade at the furthest end of the garden. Texensis varieties such as 'Gravetye Beauty'* (BOTTOM RIGHT) *are pruned similarly and, although not as showy as the Viticellas, have considerable poise and grace.* C. *'Bill Mackenzie'* (TOP RIGHT) *is a hybrid of* C. tibetana *and altogether more vigorous, its lemon-peel sepals set off by red-bronze anthers that match the stems. Flowering from midsummer, the later blooms are accompanied by attractive silvery seedheads.*

RIGHT Clematis *'Elvan' drapes* Elaeagnus *'Quicksilver' with dainty flowers and, below, purple smoke bush supports* Clematis *'Comtesse de Bouchard'.*

ABOVE *Anne Dexter particularly likes* Hydrangea involucrata *'Hortensis', the many bracts around its sterile flowers giving each floret the appearance of a tiny double begonia. Opening creamy white, the flowers age to pink, mixing pleasantly with airy astilbe and dainty lilac linaria. The curious particoloured flowers of Brazilian* Alstroemeria psittacina *lurk in the background in front of purple-leaved* Berberis thunbergii *'Somerset'.*

OPPOSITE *The richly coloured leaves of* Berberis thunbergii *'Golden Ring' make the perfect foil for a spidery flowerhead of* Eryngium amethystinum, *a particularly elegant selection of this Mediterranean species.*

is a dahlia variety which Anne Dexter believes to be about a hundred years old. Growing up to about 1.5m/5ft with semi-double rich pink flowers and dark stems, Anne Dexter calls this 'Honor Francis' after the keen gardener from Stow-on-the-Wold, Gloucestershire, who used to grow it.

The yearly cycle of maintenance in the garden can be regarded as beginning in the autumn. Then all the herbaceous plants are cut down, annuals are removed and pruning of shrubs and climbers is tackled. Anne Dexter prefers not to garden in the middle of winter, so all clearing tasks are completed before the weather becomes inclement. Once such a garden has matured, severe pruning is essential, removing virtually all the previous year's top growth to retain a balance among the plants.

In the garden of a terraced building, this is no easy task: some sixty bags of prunings have to be taken out through the house.

Shrubs such as viburnum, lonicera and privet cannot be pruned while they are draped with clematis. All summer-flowering varieties are tackled at the same time. Most of the clematis are Jackmanii or Viticella varieties which will respond well to autumn pruning, though Alpina types are left for their spring blossom. The most vigorous Viticella varieties such as 'Alba Luxurians' will grow up to 9m/30ft in a single season. Anne Dexter stresses the importance of pruning those shrubs and small trees which are destined to grow big: 'from the beginning, keep it down and if you can't keep it looking natural, prune it to a shape.' The conifers used to be regularly snipped to prevent them becoming too large, untying them each year, cutting deadwood from the centre, then trussing them up again, but are no longer subjected to such perfectionism.

The next autumn task is the protection of those tender plants which are to remain outside. The fences around the garden prevent frost from draining away, counteracting any shelter they might provide. Zone 10 plants are boarded out with a friend who has a heated greenhouse while Zone 9 plants such as *Convolvulus sabatius* (syn. *C. mauritanicus*) and *C. althaeoides* are each given a small polythene tent. It is sometimes necessary to cover the least hardy of the Zone 8 plants, including those which are intolerant of winter wet and precious plants which have just been moved or have proved unaccountably slow to get going.

When all the pruning has been completed, the borders are mulched with well-rotted farmyard manure or a proprietary organic compost. A fine grade of crushed bark is used to top dress some areas. By early winter, the most important tasks of the garden's routine maintenance have been completed and Anne Dexter can retire to the warmth of the house, though in late winter or early spring a sortie is needed to give the clematis their annual dose of fish, blood and bone fertilizer.

When the weather gets warmer, the slugs emerge from the many crevices in the dry-stone walls alongside the path. They are shown no mercy. Anne Dexter sometimes thinks she spends more on slug killers than she does on her own food and feels not the least remorse from the results of her two-monthly ministrations with a bag of slug pellets.

Throughout the summer, plants are periodically sprayed with foliar feed, incorporating a pesticide if aphids appear. Other pests and diseases have not been a problem in the past, although in 1993 there were disfiguring attacks of mildew on many of the clematis for the first time. The dense canopy of growth makes it hard for rainfall to penetrate and any soil moisture is soon lost through the large surface area of leaves. Watering is essential when the soil becomes dry, especially for clematis which are very intolerant of drought.

No one could pretend that this sort of garden is labour saving but there is no doubt it is immensely rewarding. After nearly forty years in the same garden, Anne Dexter's enthusiasm is undimmed. When many gardeners of her age would be resting on their laurels, she is to be found in the garden taking an axe to an ivy vine the thickness of a man's arm and hacking at shrubs to make space for next year's new plants. If a plant dies, it is not mourned; rather there is rejoicing at another space, another opportunity for planting. The trophies of yet more triumphant nursery visits accumulate, waiting to be planted: there is a silver buddleja, a photinia with bright red young shoots, a dogwood with brilliant winter stems, *Acer negundo* 'Flamingo' with pink-flushed white-edged leaves planned to show off yet another pale blue clematis. Where will they all go? Who knows – but space *will* be found and every plant enjoyed to the full.

BORDERS WITH TENDER PERENNIALS

Hardwick Hall

The impact of brilliant colours made by the late display of tender perennials in the West Court Borders at Hardwick Hall in Derbyshire would have pleased the original owner, the redoubtable Bess of Hardwick, Countess of Shrewsbury. She built the hall, which stands proudly on top of a hill dominating the valley below, with funds acquired through three profitable marriages. The design of 1590 is believed to be by Robert Smythson, sometimes dubbed the first great British architect, whose garden plans for palatial Thames-side mansions such as Twickenham, Ham and Wimbledon houses are of immense interest to garden historians and among the earliest accurate representations of English gardens. Sadly, little is known of the original garden at Hardwick, where only its walls with their corner summerhouses remain. In George Samuel Elgood's painting of 1897, the courtyard's east-facing border appears in late summer glory, resplendent with standard roses, clematis and Japanese anemones. This Victorian heyday is as influential in our image of Hardwick's garden as any notion of what might have existed in the days of the formidable Bess.

After Hardwick was presented to the National Trust in 1959, the borders continued for almost another thirty years to be an agreeable mix of traditional herbaceous plants with a few shrubs to give bulk. But voices of discontent came to be heard. The garden was full of interest from spring

In the sunnier of the two long side borders in the West Court in late summer, rich flower colours change gently and gradually from one group to the next. In this section, shown in the plan on pages 130–1, dahlias, chrysanthemums and penstemons are used, none of the varieties being reliably hardy in such an exposed garden in the coldest range of Zone 8. Dahlia 'Comet' (on the left) with attractive anemone-centred flowers has been used as a temporary substitute for scarce D. 'Bednall Beauty' but lacks the latter's fine bronze leaves which the design needs. Heuchera micrantha 'Palace Purple' spills over the edge. Cortaderia fulvida and Aralia elata 'Aureovariegata' are independent of the colour scheme and punctuate each of the three borders. The cortaderia imitates a fountain, not just in its shape but in the constant movement of its plumes in the slightest breeze, adding life to the planting.

127

through to early summer: bulbs and fruit blossom started the season, then the new herb garden took pride of place, followed by roses and herbaceous plants throughout June and into July; but from mid July until the end of October there was little floral display in the garden. This deficit was to be redressed by the West Court planting.

In all gardens, large or small, recent or historic, planting should respect both past traditions and present needs and likes if it is to seem right and fitting, if it is to appear to belong. For most historic gardens, delving into the past will produce plenty of themes for a design which seems utterly appropriate, making the garden look as though it has evolved over generations. At Hardwick, there

were no such themes; even the most diligent search through the archives revealed no past age of horticultural glory, no old plant lists or planting plans; all that was known was that Lady Blanche Howard laid out the planting of the West Court in 1832, including two cedars of Lebanon, one of which survives, the borders and an elaborate scheme with beds spelling out Bess's monogram, E S for Elizabeth Shrewsbury, a scheme Gertrude Jekyll deplored as 'not pretty gardening, nor particularly dignified'. The strongest criterion for planting new borders round the three outer sides of the West Court was to provide what visitors most wanted to see: a rich tapestry of colour to last when other parts of the garden were past their peak. The exuberance

and abundant colour of the nineteenth century would be essential to the new colour scheme. As Gardens Adviser responsible for Hardwick, it fell to me to produce planting plans, based on principles agreed with those reponsible for running the property.

If few precedents were to be found from Hardwick's own past, perhaps a comparison with a garden of similar size, style and date would provide a theme for the planting. The garden at Montacute in Somerset has such pronounced similarities that it might be regarded as Hardwick's twin: here the courtyard borders are similarly dominated by the powerful architecture of house and garden walls. Vita Sackville-West had provided for Montacute a planting scheme of

Cleared in early autumn ready for sterilizing (TOP), *the borders had to be cultivated to a fine and even tilth before incorporating dazomet sterilant and sealing with polythene* (ABOVE). *Applied while the ground was still warm, the dazomet had worked thoroughly in time for spring planting. Any deep-rooted weeds such as bindweed surviving the treatment were spot treated with translocated herbicide as they regrew.*

OPPOSITE *The yellow section of the border, shown overleaf on plan, has shrubs such as* Spartium junceum *and* Berberis × ottawensis *'Superba' to give height and bulk at the back. Bright yellow* Calceolaria integrifolia *at the front of the border must be raised annually from seed or overwintered under glass, and* Achillea *'Moonshine' (on the left) needs frequent replanting if it is to make the best foliage and abundant flower.*

soft colours which proved unsuccessful. It fell to Mrs Phyllis Reiss of nearby Tintinhull to provide a bolder planting scheme; her borders of red and gold leavened by stark white with sultry purple smouldering in the background can still be seen.

Here I have a confession: I admire Mrs Reiss's borders but do not entirely like them. For instance, I wonder why she chose brilliant white instead of a gentler tone, a broken white softened by a touch of buff or stone or cream. Perhaps she could not find flowers of such hues; for although we would never be satisfied with a single sort of red, blue or yellow, gardeners generally seem fairly complacent about the lack of different whites. We would never, I hope, choose the fiercest white for our garden furniture without considering more gentle tones. Why not so for flowers?

With Montacute and my reservations in mind, the parameters for the new borders at Hardwick were set. Planting plans were completed in 1985 and implemented over the following few years. Abundant rich colour was needed, lasting from midsummer until the first frosts. Few hardy plants perform for such a long period – four months or more – so a considerable proportion of tender plants would be needed to keep up the display. In their native habitats, these do not have to stop flowering and set seed before the winter; thus plants such as dahlias, mimulus, *Bidens ferulifolia* and verbenas will flower generously and continuously. There is a disadvantage: plants have to be regularly repropagated and overwintered under glass; the extra work raising these and maintaining a comparatively labour-intensive border would need to be carefully weighed and the costs set against the benefits, but their showiness and long season make this a price worth paying.

Fortunately, Hardwick is blessed with an energetic and committed team of gardeners led by Head Gardener Robin Allan. A series of propagating houses (needed also to produce

plants of suitable magnificence for the house) already existed, so such a daring scheme was feasible. An underplanting of bulbs would be a second phase in the development of the border and the wall plants would be added on completion of repairs to the garden walls and their elaborate finials.

The first stage was to clear out existing planting from one side of the courtyard, keeping in the nursery any plants that would be needed in the new design, though never 'making do' by re-using plants that were close to those specified but not so good. Some varieties on the plan were hard to find and the job of assembling them began at once. Others, like the aralias, were slow growing and also had to be acquired straight away: the border would not look complete until they had attained a reasonable size so there was no time to lose. The other two borders were to follow over the next couple of years, spreading the workload on the hard-pressed gardeners.

Meanwhile, the empty border was treated with soil sterilant. Worked into the ground in early autumn when the ground is still warm, this kills weeds and soil pathogens such as eelworm, and helps reduce the impact of rose replant disease. The surface of the ground is sealed with polythene and the sterilant works through the winter months. By mid spring, the polythene can be removed, the ground cultivated to release sterilant fumes and the border becomes ready for planting. Though such sterilants are not available to amateurs, this technique can be tackled by gardening contractors and is worth considering for more modest private gardens, particularly if rose replant disease is a problem.

As at Montacute, the repetition of plants at intervals can give unity to border design. However, if this becomes merely the repetition of a standard block a dozen times, the viewer is likely to feel cheated, and rightly so: if this were a symphony, we would not expect the composer to repeat a theme twelve times without developing it in any way; nor

LEFT *The hot colours of gloriosa daisies* (Rudbeckia hirta) *glow richly in harmony with Hardwick's honey-coloured stone in the setting sun. One of very few annuals used in the borders, the rudbeckia's late display generously repays the effort of raising plants each year from seed.*

BELOW *Hardwick's borders are relatively narrow (2.7–3m/9–10ft) to accommodate three ranks of plants, the tallest of them 2m/7ft or more high. Every bit of this width must be used to the full: there is not the luxury of a path at the back of the border to train climbers, yet to be planted to carry the display to the top of the wall. Plants such as* Bidens ferulifolia, Alchemilla mollis, Potentilla *'Gibson's Scarlet' and* Verbena *'Kemerton' sprawl across the path, increasing the effective width of the border and softening its edge. Because there is no risk of spoiling adjacent lawn, a border against a path can be set with plants right up to its margin, eliminating bare earth and making use of every scrap of available space. Rose 'Fred Loads' must be deadheaded the instant its first flush of blooms fades, reducing the new flowering stems by at least one half to encourage a second crop.*

Some plants have been less successful than others: Papaver spicatum *(syn.* P. heldreichii*), though it has flowers of a rare and charming soft apricot, produces little bloom in late summer and is inclined to look untidy, as does* Rheum palmatum *'Atrosanguineum' unless generously fed and watered;* Clerodendrum trichotomum, *a magnificent specimen surviving from the previous planting, casts such dense shade that it is hard to grow plants that continue the colour scheme beneath its branches. The bold strap-shaped leaves of* Yucca flaccida *'Ivory', less uncomfortably spiky in appearance than some of its kin, finish the border with a full stop.*

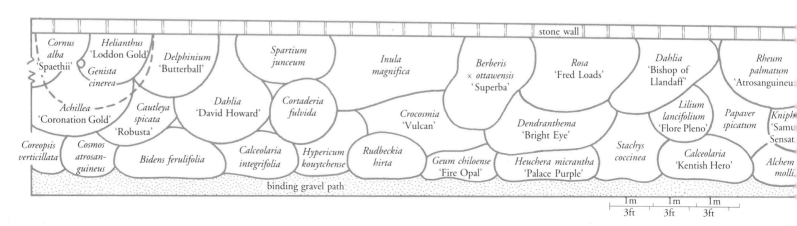

stone wall

Cornus alba 'Spaethii'

Helianthus 'Loddon Gold'

Genista cinerea

Delphinium 'Butterball'

Spartium junceum

Inula magnifica

Berberis × ottawensis 'Superba'

Rosa 'Fred Loads'

Dahlia 'Bishop of Llandaff'

Rheum palmatum 'Atrosanguineu

Achillea 'Coronation Gold'

Cautleya spicata 'Robusta'

Dahlia 'David Howard'

Cortaderia fulvida

Crocosmia 'Vulcan'

Dendranthema 'Bright Eye'

Lilium lancifolium 'Flore Pleno'

Papaver spicatum

Kniph 'Samu Sensat

Coreopsis verticillata

Cosmos atrosan- guineus

Bidens ferulifolia

Calceolaria integrifolia

Hypericum kouytchense

Rudbeckia hirta

Geum chiloense 'Fire Opal'

Heuchera micrantha 'Palace Purple'

Stachys coccinea

Calceolaria 'Kentish Hero'

Alchem molli.

binding gravel path

1m 3ft   1m 3ft   1m 3ft

130

in a palatial house would we expect to find cheaply repetitive wallpaper when rich tapestries were expected. The designer can alter the order in which plants occur in each block but even this is not enough if every part of the border is to present some new delight of plant association. At the very least some modulation of flower colour or foliage texture is needed from block to block.

At Hardwick, only the focal points provided by arching plumes of *Cortaderia fulvida* and splendidly Victorian variegated aralias are repeated. Although the toe-toe, *C. richardii*, was specified on plan, the plumes of *C. fulvida*, supplied as it so often is in error, are no less good, not so fluffy but gracefully arching and borne in midsummer, a great advantage over its South American cousin the pampas grass, *C. selloana*, which blooms too late to fulfil its required role through most of the border's season. In its youth, the cortaderia's few flower spikes, borne at divergent angles, looked untidy and faintly ridiculous; it took about five years for plants to become big enough to produce fountains of parchment plumes in scale with the scheme. The slight tenderness of these grasses is a worry but the risk is justified; no other plants could perform this function so well.

Aralias, also key plants in the border's design, are planted more rarely than they should be, because of their considerable expense and the years it takes for them to reach a telling size. Such is their unassailable

magnificence that Robin Allan is loath to prune them, though this will be necessary if they become too tall, wide or dense. It is intended that they should form an umbrella of splendid foliage at about 2m/7ft, beneath which the colour scheme of the border will flow uninterrupted.

The colours shift gradually from subfusc purples to fiery scarlet and thence to orange. These bold hues have the effect of exaggerating perspective when seen from the hall: hot reds seem close while the softer yellows, then blues at the distant end of the courtyard, appear further away than they really are. This is nothing new: Miss Jekyll described just such a progression in her *Colour Schemes for the Flower Garden* and the planting shows also our debt to her in the use of foliage colour and texture.

Some of the most generous and floriferous plants have the dullest and most amorphous foliage. This need not be a disadvantage provided other plants give textural relief: grasses such as miscanthus, rather more spiky crocosmias (especially the dazzling 'Carmin Brillant') and the dagger leaves of kniphofias each play a role, while for bold leaf shape the hostas are especially good. In the sun it is the grey-leaved sorts that look most at home, and upright *H*. 'Krossa Regal' is particularly useful where some more spreading varieties would be short of space.

Although dahlias are not blessed with the finest foliage, Miss Jekyll did not despise

them and recognized that any coarseness they might have disappears if they are incorporated into the border with other harmonious planting. Dahlias are now being rehabilitated as worthy garden plants, but they have still not entirely shaken off their image as vulgar monstrosities that belong in the kitchen garden or allotment.

There is nothing coarse about good old *D*. 'Bishop of Llandaff', its finely-cut dark leaves providing the perfect foil for rich scarlet flowers. Golden-orange 'David Howard' is also complemented by dark foliage, though not so delicate as that of the bishop. 'Bednall Beauty' is a child of the bishop and inherits his leaves, with darker flowers and a shorter habit making it useful for the front of the border.

Although bicoloured flowers can look restless, they can be useful in adding sparkle to colours such as rich purple that might otherwise disappear amid the sepulchral gloom of the appropriate section of the border; white-tipped *D*. 'Edinburgh' is ideal here. For all the dahlias used, the size of flower should be in proportion to the plant and to the scale of the border; that granted, dahlias can be perfectly respectable, even positively genteel, denizens of the border.

Other aspects of the planting are not Jekyllian. Colours are perhaps more intense and sustained over a longer period. Drifts are not used: in such narrow borders (2.7–3m/9–10ft), they would have to run almost parallel

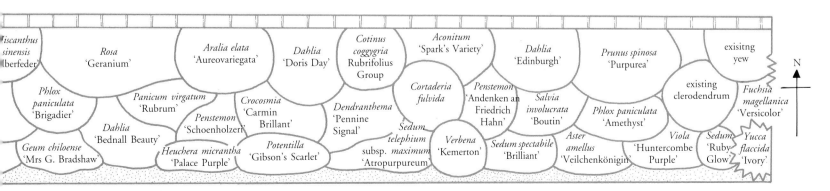

to the path and would lose much of their charm and effectiveness. Colours change little from one group of plants to the next. If colours were more mixed, such strong hues could seem unbearably frantic and the view of the entire border would appear an irritating jumble. An excess of variegated plants would also make such a scheme too hectic, so few are included. The richness of gradually changing colour and the variety of textures and forms are sufficient to carry the scheme; at close range, each eyeful provides a satisfying and harmonious picture.

Shrubs provide bulk at the back of the border, though regular pruning is needed to restrict them to the border's narrow width. For some such as *Cotinus coggygria* Rubrifolius Group and *Cornus alba* 'Spaethii', this will be tackled on a two-year cycle, each stem being

left after its first winter and pruned away during its second spring; thus in each summer, half the stems will be new and half in their second year, giving more height and a more natural habit than annually coppiced plants. Others such as the bloody nut, *Corylus maxima* 'Purpurea', will have roughly a third of the stems cut out each winter, especially those branches that stray outwards and threaten to engulf neighbouring groups. Spanish broom is difficult to prune without marring its natural habit and is best replaced whenever it becomes too large. In the border facing away from the sun, it will be particularly important to prevent shrubs becoming so large that they overshadow the shorter plants; this would give a difference in performance between the two opposite borders, planned to be identical in character

though not in planting, which could completely unbalance the scheme.

Neither the cotinus nor the purple sloe, *Prunus spinosa* 'Purpurea', is the richest colour form of its genus: it is not always necessary to have the darkest, the brightest or the most extreme form. This applies to the herbaceous planting too: though *Heuchera micrantha* 'Palace Purple' is now superseded in some respects by near black-leaved 'Bressingham Bronze', I am not sure it should be entirely supplanted by this newcomer at Hardwick.

The light brick-red flowers of a form of *Allium paniculatum* from the eminent plantswoman Valerie Finnis are distinctly subdued compared with their neighbours, burnt orange *Calceolaria* 'Kentish Hero' and red *Penstemon* 'Schoenholzeri', but this muted tone seems agreeably restrained and helps

LEFT *At the purple end of one of the two long borders, white-tipped petals of* Dahlia *'Edinburgh' enliven deep colours, complemented by the gentle tones of* Fuchsia magellanica *'Versicolor'.* Aconitum *'Spark's Variety' provides a dusky background. Like most of its clan, this likes regular replanting; selecting the fattest tubers gives long spikes and extended flowering.*

RIGHT *Though not the darkest-leaved form of this species,* Cotinus coggygria Rubrifolius Group *has foliage that is more telling for being less black and harmonizes well with* Fuchsia magellanica *'Versicolor',* Aconitum *'Spark's Variety' and* Eupatorium purpureum.

OVERLEAF Chrysanthemum *'Pennine Signal'* (LEFT) *supplies abundant colour over a long period, matching* Crocosmia *'Carmin Brillant', among the brightest and reddest of its genus.* Mimulus puniceus, Stachys coccinea *and the bronze leaves of* Dahlia *'Bishop of Llandaff'* (CENTRE) *make a subdued and subtle foil for the intense orange of* Calceolaria *'Kentish Hero'. Exotic-leaved* Cautleya *'Robusta'* (RIGHT), *seen with* Dahlia *'David Howard' behind, is perfectly hardy in Zone 8, as is the chocolate dahlia* (Cosmos atrosanguineus) *in front, provided its roots are not frozen. The flowers of the cosmos match the mahogany red bracts of the cautleya.*

132

pacify a potential clash between disputatious colours. Tender *Stachys coccinea*, in spite of its name, is a similarly bricky salmon and equally irenic among reds and vermilions which can be far from restful.

Sprawlers and weavers are useful for the front of the border, softening the hard edge and uniting neighbouring groups. Cranesbills will fill this role for muted colour schemes but at Hardwick the job is done by verbenas such as magenta 'Kemerton' and the versatile *Bidens ferulifolia*, as invaluable in the border as it is in containers.

Though they can seldom be persuaded to provide a second autumn crop of flowers, it is hard to exclude delphiniums from grand borders; however, the most magnificent cultivars can be overpowering in smaller gardens. At Hardwick, the brilliant gentian

blue of *D.* 'Fenella' is so dominant and unbalancing that I find myself wanting either more delphiniums at intervals throughout the backs of the borders, repeating the strong vertical accent, or none at all. *D.* 'Butterball', included to provide a gentle contrast toning up the surrounding blues, has proved a poor doer and should perhaps be replaced by new 'Sunkissed', reputedly more vigorous and reliable, or even a verbascum.

Where magnificence and opulence are required, lilies will be needed, planted in generous quantities. To be sure of healthy, virus-free stock, the largest-sized bulbs should be chosen; these will also have most flowers per stem, ensuring that their display lasts for as long as possible. True, not all of them will perform indefinitely; most will diminish in size and number and will need 'topping up'

from year to year, or even replacing entirely if traces of virus are seen. Dr North's triploid Scottish hybrids are particularly persistent and reliable, though unaccountably hard to find: 'Orestes', originally specified on plan, has had to be replaced by an Asiatic hybrid. Soft orange double tiger lily, *Lilium lancifolium* 'Flore Pleno' and *L. henryi* are also used at Hardwick.

The magenta flowers of tender *Salvia involucrata* 'Boutin' are the more valuable for being borne about three weeks before the more common variety, 'Bethellii'. *Argyranthemum* 'Jamaica Primrose' is another star performer, again propagated annually from cuttings which it is invariably reluctant to produce; the oft-quoted rule that non-flowering stems should be chosen for cuttings of this genus should be treated with some

caution as this can lead to the selection of variants that will produce flowers only when days are less than twelve hours long.

Penstemons are immensely valuable in providing flower from midsummer to the frosts but at Hardwick it is always necessary to take cuttings in late summer as a precaution against winter losses. 'Andenken an Friedrich Hahn' ('Garnet') and 'Schoenholzeri' are tough and eternally reliable, though 'Alice Hindley' and 'Papal Purple' are less frost-hardy. Some of the plummy-coloured varieties, 'Burgundy', 'Port Wine' and 'Rich Ruby', are wonderful but might disappear among dusky foliage if used in the purple-to-red sections of the borders.

It is curious that so few gardeners plant for late summer and autumn. As though exhausted by the annual horticultural frenzy of spring and early summer, they seem all too willing to surrender to the inevitable onset of winter months before they need do so. The West Court Borders show what they might be missing: neither the scale nor the character of this planting is so grand that it could not easily be adapted for more modest gardens.

Because of this reluctance to garden for autumn, chrysanthemums are seldom used to good effect, curious since they were so popular as recently as the 1950s. Korean varieties are found in a few National Trust gardens, notably at Sissinghurst and in the Autumn Border at Powis Castle, but some excellent recent varieties such as 'Pennine Signal', used here, are unaccountably neglected.

Roses are included to give bulk and continual flower, including tall varieties like 'Chinatown' and 'Fred Loads' which will grade into the fairly steeply banked borders, with shorter 'Marlena' to the front; perhaps 'The Times' would be better here, its wonderfully rich red blooms set off by the darkest foliage. Hydrangeas, intended to provide pure blue flowers in shade beneath the cedars, have resolutely refused to be anything other than bright pink despite the administration of alum; they look appalling and will have to be replaced.

At Hardwick, it was intended that each plant should produce an agreeable effect with every other group it adjoins; this is of course a counsel of perfection. The borders were not to be a series of groups of, say, three plants that go well together followed by another three. If each plant is in harmonious relationship with every one of its neighbours, the border becomes a continual progression of delightful associations rather than a series of incidents. At Anglesey Abbey near Cambridge, the borders are unsurpassed for grand scale and impeccable cultivation, set off by some of the finest emerald turf in the land, but, as magnificent as they are, they might be even more impressive if every plant association were contrived for optimum effect.

For any border to last in glory for a number of months, the proportion of groups in flower must always be at least one third if the interplay of colours is to 'read' from group to group along its length; for the most gorgeous displays, the proportion should be at least one half. Even two adjacent groups showing nothing but green foliage or a single oversized group that is not performing can disrupt the progression of interrelated colours.

The photographer can, and does, overcome such inadequacies in design by avoiding gaps and large blocks of green, positioning the camera where these lie outside the frame or where they disappear behind some floral marvel in the appropriate colour; but such dodges are not available to the visitor, who must walk past these lacunae and be disappointed, albeit subconsciously, by the border's shortcomings. At Hardwick, the proportion of plant groups performing to

those past or before their peak is more than one half during each of the late summer and autumn months, so such lapses are rare.

A good deal remains to be done at Hardwick before the borders are complete. There is no reason why the borders should not provide an equally dazzling display in the spring, though bulbs have yet to be added. The colour scheme for this could be quite different but will perhaps follow the same progression as the herbaceous planting, using tulips, narcissi, hyacinths and muscari. The tulips will probably be the most important of these, their foliage remaining much more tidy than that of narcissi after flowering. Wall plants will also be needed so that the colour scheme extends to the top of the garden walls; clematis will doubtless predominate, planted where they can also drape themselves over some of the shrubs. The best borders are never created simply by a few days' planning with pencil and paper. Years of constant reworking and refinement are needed if every plant is to play its role to the full. Even when near perfection is reached, the planting must change from year to year if it is to be fresh and stagnation avoided.

The use of so many hues, almost every colour in the rainbow, and such intensity of display will not appeal to all. Those who prefer more gentle and restrained use of colour might agree with Graham Thomas in finding such ostentation repellent. But, though now dimmed by the passing years, Hardwick was built for show, to dazzle all who saw it by the virtuosity of its architecture, the splendour of its furnishings, the rich and varied colour of its tapestries and wall hangings. Surely Bess of Hardwick would have wanted her garden to be as ostentatious as her house. Those who find the vibrancy of the borders too much can visit in May or June before the peak of the display. There are plenty of other gardens where colours are quieter and less intense; there are few like Hardwick. *Vive la différence!*

OPPOSITE *The shortest of the three borders, running across the furthest end of the West Court and divided in two by a central gatehouse, is planted predominantly with blue and white contrasted with a little soft yellow. Blue, the colour of the hazy distant summer view, exaggerates perspective, making the borders seem even further away from the house than they really are. Dainty* Dicentra *'Langtrees' fronts the border next to glaucous* Hosta fortunei *var.* hyacinthina *and* Nepeta *'Six Hills Giant', framing a seat painted a gentle broken white. Golden aralia dominates the foreground while fierce blue* Delphinium *'Fenella' draws the eye towards the gazebo at the corner of the court.*

ABOVE *Harmonies of green and gold in the sunnier of the two long borders feature* Hosta montana *'Aureomarginata', among the most dazzling of the plantain lilies and one of few variegated plants used here. At Hardwick it is reasonably tolerant of summer sun, though in the hottest, driest seasons its bright edges can become scorched as autumn approaches. Rose 'Chinatown', a Floribunda so vigorous that it is sometimes classed as a Shrub, has buttery yellow flowers that soften the slightly acid tones of the hosta and chartreuse* Alchemilla mollis. *Fine, even down on the alchemilla leaves holds glistening droplets of dew; for good late summer and autumn foliage, it should be cut back after flowering to grow again.*

137

Gardens to visit

With the exception of Le Pontrancart, all the gardens in *Best Borders* may be visited by the general public. Opening arrangements, for these and other gardens in the British Isles with notable borders or beds listed below, are given in *The Good Gardens Guide* (ed. G. Rose and P. King), published annually by Vermilion, London.

ENGLAND

Anglesey Abbey, Lode, Cambridgeshire
Audley End, Saffron Walden, Essex
Barnsley House, near Cirencester, Gloucestershire
Barrington Court, Barrington, Somerset
Beningbrough Hall, near York
Beth Chatto Gardens, Elmstead Market, Essex
Blickling Hall, Aylsham, Norfolk
Bramdean House, near Alresford, Hampshire
Bressingham Gardens, Diss, Norfolk
Broughton Castle, near Banbury, Oxfordshire
Burford House Gardens, Tenbury Wells, Hereford and Worcester
Buscot Park, Faringdon, Oxfordshire
Clare College, Cambridge
Hadspen Garden, Castle Cary, Somerset
Hatfield House, Hatfield, Hertfordshire
Jenkyn Place, Bentley, Hampshire
Kiftsgate Court, Chipping Campden, Gloucestershire
Mottisfont Abbey Garden, near Romsey, Hampshire
Newby Hall, Ripon, North Yorkshire
Polesden Lacey, Great Bookham, Surrey
Royal Horticultural Society's Garden, Wisley, Surrey
Tintinhull House Garden, near Yeovil, Somerset
Wallington, Cambo, Northumberland

SCOTLAND

Castle of Mey, Caithness, Highlands
Crathes Castle, Banchory, Grampian
Royal Botanic Garden, Edinburgh

WALES

Powis Castle, Welshpool, Powys

IRELAND

Mount Stewart, Newtownards, Co. Down
45 Sandford Road, Ranelagh, Dublin
The Shackleton Garden, Clonsilla, Co. Dublin

CANADA

Van Dusen Botanic Garden, Vancouver, British Columbia

UNITED STATES

Chicago Botanic Garden, Lake Cook Road, Glencoe, IL 60022
The Conservatory Garden, Central Park, 105th St and 5th Avenue, New York, NY 10003
Dumbarton Oaks, 1703 32nd Street NW, Washington, DC 20007
Longwood Gardens, Route 1, Kennett Square, PA 19348
North Carolina Botanical Garden, University of North Carolina, Chapel Hill, NC 27514
Old Westbury Gardens, 71 Old Westbury Road, Old Westbury, NY 11568
Stonecrop, Cold Spring, NY 10516 (Visits by appointment, Monday, Wednesday, Friday, telephone 914-265-2000)
Wave Hill, 675 West 252nd Street, Bronx, NY 10471
Western Hills Nursery, 16250 Coleman Valley Road, Occidental, CA 95465

Further reading

The titles listed below cover historical sources, including Victorian and Edwardian paintings of borders in *Some English Gardens* and *Painted Gardens* (some showing gardens featured in *Best Borders*), plus useful works describing a wide range of border plants and their cultivation.

Armitage, Allan M. *Herbaceous Perennial Plants*, Varsity Press, Athens, Georgia, 1989
Balmori, Diana, McGuire, Diane Kostial & McPeck, Eleanor M. *Beatrix Farrand's American Landscapes*, Sagapress, Portland, Oregon, 1985
Bisgrove, Richard. *The Gardens of Gertrude Jekyll*, Frances Lincoln, London, 1992
Bloom, Alan. *Perennials in Island Beds*, Faber & Faber, London, 1987
Brown, Jane. *Eminent Gardeners*, Viking, London, 1990
Brown, Jane. *Lanning Roper and his Gardens*, Weidenfeld & Nicolson, London, 1987
Clarke, Ethne. *Hidcote*, Michael Joseph, London, 1989
Clausen, Ruth Rogers & Ekstrom, Nicolas H. *Perennials for American Gardens*, Random House, New York, 1989
Elliott, Brent. *Victorian Gardens*, Batsford, London, 1986
Hobhouse, Penelope. *Borders*, Pavilion, London, 1989
Hobhouse, Penelope. *Colour in Your Garden*, Frances Lincoln, London, 1985
Hobhouse, Penelope & Wood, Christopher. *Painted Gardens*, Pavilion, London, 1988
Jekyll, Gertrude. *Colour Schemes for the Flower Garden*, Frances Lincoln, 1987
Jekyll, Gertrude & Elgood, George Samuel. *Some English Gardens*, London, 1904
Jekyll, Gertrude. *Wood and Garden*, Longman, London, 1899
Jelitto, Leo & Schacht, Wilhelm. *Hardy Herbaceous Perennials* (3rd ed.), Batsford, London, 1990
Jellicoe, Geoffrey & Susan, Goode, Patrick & Lancaster, Michael. *The Oxford Companion to Gardens*, Oxford University Press, 1986
Keen, Mary. *The Garden Border Book*, Viking, London, 1987
Laird, Mark. '"Our Equally Favourite Hobby Horse": The Flower Gardens of Lady Elizabeth Lee at Hartwell and the 2nd Earl Harcourt at Nuneham Courtenay', *Journal of the Garden History Society*, Vol 18, No 2 pp103-54, 1990
Lloyd, Christopher. *Foliage Plants*, Collins, London, 1973
Lloyd, Christopher & Rice, Graham. *Garden Flowers from Seed*, Viking, London, 1991
Lloyd, Christopher. *The Mixed Border*, Collingridge, London, 1957
Lloyd, Christopher. *The Mixed Border*, Wisley Handbook Series, Royal Horticultural Society, London, 1986
Lloyd, Christopher. *The Well-Tempered Garden*, Collins, London, 1970
Loudon, John Claudius. *Encyclopaedia of Gardening*, Longman, London, 1822
Lovejoy, Anne. *American Mixed Border*, Macmillan, New York, 1993
Miller, Philip. *Gardeners Dictionary*, London, 1731
M'Intosh, Charles. *The Flower Garden*, London, 1838
Robinson, William. *The English Flower Garden*, John Murray, London, 1883
Robinson, William. *Hardy Flowers*, Warne, London, 1871
Scott-James, Anne. *Sissinghurst*, Michael Joseph, London, 1974
Taylor, Jane. *The Milder Garden*, Dent, London, 1990
Thomas, Graham Stuart. *The Art of Planting*, Dent, London, 1984
Thomas, Graham Stuart. *Perennial Garden Plants* (3rd ed.), Dent, London, 1990
Underwood, Mrs Desmond. *Grey and Silver Plants*, Collins, London, 1971
Verey, Rosemary. *Good Planting*, Frances Lincoln, London, 1990

Index

Page numbers in *italics* refer to illustrations and planting plans. Page numbers in **bold** refer to principal references, including illustrations. For an explanation of hardiness zones, e.g. **Z7**, see end of index.

A

Acanthus spinosus **Z7** *99*

Acer 119; *A. negundo* 'Flamingo' **Z3** 125; *A.n.* 'Variegatum' **Z3** (variegated box elder) *119*, *123*; *A. platanoides* 'Crimson King' **Z4** *45*, 48

Achillea 7, 25, *25*, 31, 54, 56, 80; *A.* 'Anthea' **Z5** 25; *A.* 'Gold Plate' **Z4** *80*, *81*, *88*, 89; *A. millefolium* 'Cerise Queen' **Z3** *32*, 35; *A.* 'Moonshine' **Z4** 25, *129*; *A. ptarmica* 'The Pearl' **Z4** *38*, 56, 77, 80, 82; *A.* 'Taygetea' **Z5** *4*

Aconitum 17; *A.* × *cammarum* 'Bicolor' **Z4** *79*; *A.* 'Spark's Variety' **Z4** *32*, 45, *47*, 48, *49*, *131*, *132*

Actaea alba (doll's eyes) **Z4** *119*, 121

Acton, Tommy 20, 25

Adams, Frank (d.1939) 46

Adiantum 119; *A. capillus-veneris* **Z8** *119*

aeration, turf 53

Ageratum 57; *A.* 'Tall Blue' *38*, *43*

Ajuga reptans 'Atropurpurea' **Z3** *49*

Alchemilla mollis **Z4** *10*, 99, 103, *108*, *130*, *137*

Allan, Robin 129, 131

Allium 115; *A. cernuum* **Z3** 12, *108*; *A. christophii* (syn. *A. albopilosum*) **Z4** 15, *98*, *98*, *99*, *103*; *A. giganteum* **Z6** 15, *21*, 25, *27*; *A. neapolitanum* **Z6** 102; *A. paniculatum* **Z1** *132*; *A. sphaerocephalon* **Z6** *63*, 64, *65*, *67*; *A. stipitatum* **Z4** 64, *67*

alpines 120

Alstroemeria 15, 25, 102; *A.* Ligtu hybrid **Z6** 64, *99*, 102; *A. psittacina* **Z8** *124*

Alyogyne hakeifolia **Z10** *67*, 73

Amaranthus 93; *A.* 'Red Fox' *88*

Anaphalis margaritacea var. *cinnamomea* **Z4** *32*

Anchusa 34, 68; *A.* 'Loddon Royalist' **Z4** 64

Anemone 6, *54*; *A.* × *hybrida* **Z5** *98*; *A.* × *h.* 'Honorine Jobert' *79*, *99*; Japanese a. 102, 127

Anemonopsis **Z5** 121

Anglesey Abbey, Cambridgeshire *9*, 136

annuals 7, 11, 12, *12*, **36-43**, **54-61**, 123

Anthemis tinctoria **Z4** *25*, 34; *A.t.* 'E.C. Buxton' *25*, *81*; *A.t.* 'Grallach Gold'

32, 34, *35*; *A.t.* 'Kelwayi' *27*

Antirrhinum 11, *38*, 40, 41, *54*, 56, *56*, *57*, 58, 60, 61, *61*, *88*, 93; *A.* 'Coronette Crimson' *38*; *A.* 'Maryland Red' *38*; *A.* 'Sonnet Pink' *38*

aphids 40, 125

Aquilegia 73

Aralia elata 'Aureovariegata' **Z4** *127*, 129, *131*, 131, 137

Argyranthemum (syn. *Chrysanthemum*) **Z9** 12, 30; *A.* 'Jamaica Primrose' *32*, 34, *35*, 132

Arley Hall, Cheshire: Double Borders 6, 11, **14-27**

Artemisia lactiflora **Z4** *79*, *81*, *99*; *A.l.* 'Guizhou' *86*; *A. ludoviciana* **Z4** 25, *54*, *57*

Aruncus dioicus **Z3** *81*, 82

Arundo donax **Z8** 105

Asarum europaeum **Z4** *119*

Ash, G. Baron 29

Ashbrook, Viscountess 15, 20, 21, 24

Asplenium scolopendrium SEE *Phyllitis s.*

Aster 10, 56, 57, 64, 73, 74, *74*, 98; *A. amellus* 'Veilchenkönigin' (*A.a.* 'Violet Queen') **Z5** *131*; *A.* × *frikartii* **Z5** *79*; *A.* × *f.* 'Mönch' *88*; *A.* × *f.* 'Wunder von Stäfa' (*A.* × *f.* 'Wonder of Stäfa') *67*, 74; *A. novi-belgii* **Z4** *32*, 64, 68; *A.n.-b.* 'Carnival' *67*; *A.n.-b.* 'Cliff Lewis' *65*, *67*, *74*; *A.n.-b.* 'Climax' 42, *67*; *A. sedifolius* (syn. *A. acris*) **Z4** *104*; *A. tradescantii* **Z3** 31, *32*; *A. turbinellus* **Z3** 74

aster, China 39

Astilbe 79, 119, *124*; *A. simplicifolia* **Z4** 119

Astrantia major **Z4** 24, *27*; *A.m. rubra* 24; *A. maxima* **Z4** 24, *99*

Atriplex hortensis var. *rubra* (purple orach) 15, *27*, *88*

Aucuba japonica 'Variegata' **Z7** *119* auricula 6, 7

B

Ballota pseudodictamnus **Z7** *27*

balsam (*Impatiens balsamifera*) 58, *61*

Baptisia australis **Z3** *67*

bark, crushed 124

Bateman, James (1811-97) 16

beech hedges 20, 109

beetle, lily *80*, 83

beetroot, ornamental 93

Begonia 46, 52; *B.* 'Hatton Castle' 52

bellflower SEE *Campanula*

Bennett, Walter 46

Berberis dictyophylla **Z6** 46; *B.* × *ottawensis* 'Superba' **Z5** 103, *129*, *130*; *B. thunbergii* **Z5** 120; *B.t.* f.

atropurpurea 110; *B.t.* 'Atropurpurea Nana' *16*, *17*, 25, *25*; *B.t.* 'Golden Ring' 120, *124*; *B.t.* 'Rose Glow' 120; *B.t.* 'Somerset' 120, *124*

Bergenia 10, 77, 78, 98, *119*; *B.* 'Ballawley' **Z4** *119*; *B. ciliata* f. *ligulata* **Z5** *119*; *B. cordifolia* 'Purpurea' **Z3** *81*

Bess of Hardwick (Elizabeth, Countess of Shrewsbury) (1518-1608) 127, 128, 137

Bidens ferulifolia **Z9** 129, *130*, 132

biennials *12*

bindweed 42, 83, *129*

biological control of pests 40

blackfly 74

blackspot 74

Blechnum 119; *B. spicant* **Z4** *119*; *B.tabulare* **Z8** *119*

Bloom, Alan *9*, 10

blue lace flower SEE *Trachymene*

borders, herbaceous 6, 85; mixed b. 85; b. orientation *38*

box 120; b. edging 7; b. hedges 20; b. scrollwork 7; topiary b. *7*

box elder SEE *Acer negundo*

Branch, Arthur 120

Brazier, Charles 86

Bressingham, Norfolk *9*, 10

broom, Mount Etna (*Genista aetnensis*) **Z8** 98; Spanish b. (*Spartium junceum*) **Z8** 132

Broughton Castle, Oxfordshire 12, *12*

Brown, Lancelot 'Capability' (1716-83) 16

browntop 53

brushwood 30, 31, 34, 41, *41*, 72, *72*, 73

Buddleja 38, 48, *49*, 102, *108*, 125; *B. davidii* **Z6** *38*; *B.d.* 'Royal Red' 45; *B.* 'Lochinch' **Z7** *99*, *105*

bulbs, spring *103*, 129, 137

Buphthalmum salicifolium **Z4** *81*, 85

Burrows, George Harry 46, 47, 52

Buxus (box) 120; *B. microphylla* **Z5** 20; *B. sempervirens* **Z6** 20; *B.s.* 'Handsworthiensis' 20; *B.s.* 'Hardwickensis' 20; *B.s.* 'Notata' (syn. *B.s.* 'Gold Tip') 20

C

Cabot, Frank 12

Caiophora lateritia 120

Calceolaria integrifolia **Z9** *129*, *130*; *C.* 'Kentish Hero' **Z10** *130*, 132, *132*

Campanula (bellflower) *12*, 25, 115, 120, *121*; *C.* 'Elizabeth' **Z4** 115; *C. glomerata* **Z3** *108*; *C.g.* 'Superba' 68; *C. lactiflora* **Z4** 21, 25, *27*, 64, *67*,

109; *C. latiloba* 'Highcliffe Variety' **Z4** 115; *C. persicifolia* **Z4** *27*; *C.p.* double white *119*; *C. primulifolia* **Z8** 119; *C. trachelium* 'Bernice' **Z4** *119*

campion, double rose 6

Canna 11, 37, *38*, 41, 42, 52, *88*, 93, 94, 102; *C. indica* 'Purpurea' **Z9** *38*, *39*; *C.* 'Roi Humbert' **Z9** 45, *49*, 52

capsid 74

Caragana arborescens 'Lorbergii' **Z3** 94

cardoon SEE *Cynara cardunculus*

Carex 119; *C. conica* 'Snowline' **Z7** *119*

carnation 7

Caryopteris 74

castor oil plant SEE *Ricinus*

catmint SEE *Nepeta*

Cautleya spicata 'Robusta' **Z7** *130*, 132

Ceanothus 21; *C.* 'Gloire de Versailles' **Z7** *27*, *88*, 89

cedar of Lebanon **Z6** *47*, 128

Centaurea 'Black Knight' *88*; *C. hypoleuca* 'John Coutts' **Z4** *108*, *110*; *C. macrocephala* **Z3** *32*, 35

Centranthus ruber **Z5** *32*

Cephalaria gigantea **Z3** 17

Chamaecyparis lawsoniana **Z6** 119

chard, ruby *88*, 93

Chatto, Beth 111

Cheiranthus SEE *Erysimum*

Choisya ternata **Z8** 119

chrysanthemum (SEE ALSO *Argyranthemum, Dendranthema*) 41, *127*, 136; c. 'Bright Eye' **Z5** *130*; Korean c. **Z5** 136; c. 'Pennine Signal' **Z5** *131*, *132*, 136

Chrysosplenium oppositifolium **Z5** 119

Cimicifuga 99; *C. simplex* Atropurpurea Group **Z4** *27*, 49

Clematis 25, *38*, 41, **68-69**, 74, *74*, 82, *99*, 103, *105*, 117, *118*, 120, 123, *123*, 124, 125, 127, 137; *C.* 'Abundance' **Z6** *67*; *C.* 'Alba Luxurians' **Z6** 95, *123*, 124; Alpina c. **Z5** 124; *C.* 'Bill Mackenzie' **Z6** *123*; *C.* × *durandii* **Z6** *67*, 72, 120; *C.* 'Elvan' **Z6** 120; *C.* 'Ernest Markham' **Z6** *67*; *C.* 'Etoile Violette' **Z6** 49, *67*, 68, 74; *C.* 'Gipsy Queen' **Z6** 49; *C.* 'Gravetye Beauty' **Z6** *123*; *C. heracleifolia* **Z5** *108*; herbaceous c. 120; *C. integrifolia* var. *albiflora* **Z3** 120; *C.i.* 'Pastel Blue' **Z3** 120; *C.i.* 'Rosea' **Z3** 120; *C.* × *jackmanii* **Z5** 98; Jackmanii c. **Z6** 68, 124; *C.* 'Jackmanii Alba' **Z6** 103; *C.* 'Jackmanii Superba' **Z6** 103; *C.* × *jouiniana* 'Praecox' **Z5** 65, *67*, 69; *C.* 'Kermesina' **Z6** *88*, 89; *C.* 'Leonidas' **Z6** 69; *C.* 'Mme Julia

Correvon' **Z6** *69*; *C.* 'Minuet' **Z6**
123; *C.* 'Niobe' **Z6** *67*; *C.* 'Perle
d'Azur' **Z6** *65, 67, 68, 74, 88, 89,* 94;
C. recta **Z3** *7*; *C.r.* 'Purpurea' *103*; *C.*
'Royal Velours' **Z6** *88, 89*; *C.* 'Star of
India' **Z6** *68*; Texensis c. **Z6** *123*; *C.
tibetana* **Z6** *123*; *C.* 'Victoria' **Z6** 69;
C. 'Ville de Lyon' **Z6** *43, 68*; *C.
viticella* 'Purpurea Plena Elegans' **Z6**
123; Viticella c. **Z6** *68, 123*, 124
clematis wilt 69
Cleome 38, 41, 58; *C. hassleriana* 'Rose
Queen' *43*; *C.h.* 'Violet Queen' *67*
Clerodendrum trichotomum **Z7** *118, 119,
130, 131*
climbers 117, *130*, 137
Cliveden, Buckinghamshire 30, **76-85**
Comber, James (*c*1866-1953) 37
Compositae *43*
compost 74, 93, 124; mushroom c. *42,
93, 102*; seed and potting c. 39, 40, 52
coneflower SEE *Rudbeckia*
conifers 120, 124
Convolvulus althaeoides **Z9** 124; *C.
sabatius* (syn. *C. mauritanicus*) **Z9**
117, 124; *C. tricolor*, *C.t.* 'Dark Blue'
43
Cook, Sarah 69, 73, 74
Cordyline australis **Z9** *47, 49, 52*
Coreopsis 61; *C. verticillata* **Z4** *32*
cornflower 93
Cornus (dogwood) 125; *C. alba* 'Aurea'
Z2 120; *C.a.* 'Elegantissima' **Z2** *54*;
C.a. 'Spaethii' **Z2** *88*, 94, 120, 132; *C.
kousa* **Z5** *41*
Corrin, Fred 31, 34, 35
Cortaderia fulvida **Z8** *127, 130, 131,
131*; *C. richardii* (toe-toe) **Z8** 131; *C.
selloana* (pampas grass) **Z7** 131
Corylus maxima 'Purpurea' (purple filbert
or hazel, bloody nut) **Z5** 48, *49*, 98,
132
Cosmos 39, *43*, 54, *54, 56, 56, 57,* 58,
61, 93; *C. atrosanguineus* (chocolate
dahlia) **Z8** 52, *132*; *C. bipinnatus* 'Sea
Shells' *38*; *C.b.* 'Sensation' 102; *C.b.*
'Sonata' *123*
Cotinus coggygria (smoke bush) **Z5** *7, 63,
64, 97, 99, 103*; *C.c.* 'Foliis Purpureis'
67, 120; *C.c.* 'Royal Purple' *49, 98,
105*; *C.c.* Rubrifolius Group *131, 132,
132*
Cotton, Philip 82, *83, 85*
Crambe cordifolia **Z6** *12, 79, 85*
cranesbill SEE *Geranium*
Crathes Castle, Grampian, Scotland 83
Cresson, Charles *10*
Crinum × powellii 'Album' **Z7** 56, *56, 57*
Crocosmia 17, 25, *25, 27*, 80, 85, 88, 89,

94, 131; *C.* 'Carmin Brillant' **Z6** *131,
131, 132*; *C. × crocosmiiflora* **Z5** *17,
27*; *C.* 'Lucifer' **Z5** *15, 21, 25, 27, 49,
88, 99*; *C. masoniorum* **Z7** *81, 88*; *C.
paniculata* **Z6** *81*; *C.* 'Star of the East'
Z6 *88*; *C.* 'Vulcan' **Z5** *130*
Crocus 6, 120
crown imperial SEE *Fritillaria imperialis*
Cynara cardunculus (cardoon) **Z7** *10, 67,
68, 74, 98, 99, 102, 107, 108, 114,
115*
cypress, Leyland **Z7** 20

D
Dactylorhiza maculata **Z5** *119*
daffodil 6
Dahlia **Z9** *4, 11, 17, 25, 27,* 30, 34, *38,
41, 42, 43, 52, 57, 59, 61, 64, 65, 72,*
88, 89, 93, 94, 95, *95,* 102, *108,* 110,
127, 129, 131; *D.* 'Arabian Night'
110; *D.* 'Australia Red' *38*; *D.*
'Bednall Beauty' *127, 130,* 131; *D.*
'Bishop of Llandaff' 25, *47, 52, 88,
93, 94, 130,* 131, *132*; *D.* 'Blaisdon
Red' *4, 88, 93, 94*; *D.* 'Bloodstone'
47, 49, 52; *D.* 'Comet' *127*; *D.* 'David
Howard' *37, 130,* 131, *132*; *D.* 'Doris
Day' *49, 52, 131*; *D.* 'Edinburgh' *63,
64, 131,* 131, *132*; *D.* 'Gaiety' *27*; *D.*
'Good Intent' *67*; *D.* 'Greenside
Antonia' *43*; *D.* 'Grenadier' *49, 52,
94*; *D.* 'Honor Francis' *117*, 124; *D.*
'Pink Michigan' *65, 67*; *D.* 'Requiem'
65, 67; *D.* 'Rothesay Castle' *43*; *D.*
'Symbol' *43*; *D.* 'Tally Ho' 25; *D.*
'Zonnegoud' *38*
dahlia, chocolate, SEE *Cosmos
atrosanguineus*
daisy, gloriosa (*Rudbeckis hirta*) *130*
daisy, Shasta **Z4** *77, 82*
damping off 40
daylily SEE *Hemerocallis*
dazomet soil sterilant *129*
deadheading *59*
Deinanthe bifida **Z5** *119*
Delphinium 12, *17, 25, 27, 30, 31, 34,*
59, 68, 73, 80, 81, 83, 85, 89, 102,
107, 108, 115, 132; *D.* Black Knight
Group **Z2** *48, 49, 67, 72*; *D.*
Blackmore and Langdon's hybrids **Z2**
67; *D.* 'Butterball' **Z2** *130,* 132; *D.*
'Fenella' **Z2** 132, *137*; *D.* 'Sunkissed'
Z2 132
Dendranthema (chrysanthemum) 'Bright
Eye' **Z5** *130*; *D.* 'Pennine Signal' **Z5**
131, 132
Deramore, Nina, Lady 107, 108
Deutzia 98
Dexter, Mrs Anne, garden in Oxford 12,

116-125
Dianthus amurensis **Z3** *67*; *D.* 'Annabel'
Z4 *67*
Diascia 118, 121; *D. rigescens* **Z8** *27*; *D.
vigilis* **Z8** 121
Dicentra 123; *D.* 'Langtrees' **Z4** *137*; *D.
scandens* **Z6** 120
Dierama pulcherrimum **Z7** *67*
Digitalis 'Sutton's Apricot' **Z3** *103*
dogwood SEE *Cornus*
doll's eyes SEE *Actaea alba*
Douglas, David (1799-1834) 8
drifts 56, *56,* 78, 79, *80*, 131
Dryopteris 119; *D. affinis* 'Crispa
Congesta' **Z4** *119*; *D. erythrosora* **Z5**
119
dusty miller SEE *Senecio cineraria*

E
Echinacea 89
Echinops 21, 25, 34, *54, 56, 56, 89*; *E.
bannaticus* **Z3** *27*; *E.b.* 'Blue Globe'
32, 57; *E. ritro* **Z3** *38, 88*
eelworm, stem 42
Egerton-Warburton, Piers 20
Egerton-Warburton, Rowland (1804-91)
16
Elaeagnus 119, 120; *E. × ebbingei*
'Limelight' **Z7** *119*; *E.* 'Quicksilver'
Z3 *118,* 120
elder SEE *Sambucus*
Elgood, George Samuel (1851-1943) 20,
37, 127
elm, Dickson's golden SEE *Ulmus*
'Dicksonii'
Elymus; *E. hispidus* **Z5** *57*; *E.
magellanicus* (syns. *Agropyron
magellanicum*, *A. pubiflorum*) **Z8** *108,
109*, 114
Emes, William (1730-1803) 16
Epimedium 121; *E. grandiflorum*
'Nanum' **Z5** *119*
Eremurus robustus **Z5** *98*
Erigeron 32
Erodium 120
Eryngium 56; *E. amethystinum* **Z3** *124*;
E. bourgatii 'Oxford Blue' **Z5** *121*; *E.
pandanifolium* **Z8-9** *98, 102*; *E.
× tripartitum* **Z5** *54, 63, 67, 68*
Erysimum 69, 72; *E.* 'Bowles' Mauve' **Z8**
69; *E.* 'Constant Cheer' **Z7** *67*; *E.*
'Mrs L.K. Elmhirst' **Z7** 69; *E. cheiri*
(syn. *Cheiranthus c.*) 'Ruby Gem' (syn.
E.c. 'Purple Queen') **Z7** 69
Eucalyptus 99, 102
Euonymus 118, 120; *E. fortunei* **Z5** 120;
E.f. 'Silver Queen' 102
Eupatorium purpureum **Z3** *38, 43, 132*
Euphorbia characias subsp. *wulfenii*

'Lambrook Gold' **Z7** *119*; *E.
polychroma* **Z4** *81*; *E. schillingii* **Z7** *4*;
E. sikkimensis **Z6** *118*
evergreens 120

F
Fatsia **Z8** 120
ferns *119*, 121
fertilizer 34, 39, 41, 48, 74, 102, 125
fescue, Chewing's 53
filbert SEE *Corylus maxima*
Filipendula palmata **Z3** *27*; *F. rubra*
'Venusta' **Z3** *29, 32,* 34
Finnis, Valerie 132
Flower, The Hon. Michael 27
Foerster, Karl (1874-1970) 10
foliage, gold 120; grey or silver f. *16, 63,
64*, 78, 111, 114; maroon f. 120;
variegated f. 115, 120, 132
forget-me-not *103*
Foster, Mr and the Hon. Mrs Charles 25
foxglove **Z3** *109*
frets 6
Fritillaria (fritillary) 6; *F. imperialis*
(crown imperial) **Z5** 6, *7, 80*; *F.
pyrenaica* **Z6** 121
Fuchsia 46; *F.* 'Koralle' **Z10** *49*; *F.
magellanica* 'Versicolor' **Z8** *131, 132*;
Triphylla f. **Z10** 52
fungicide 39, 40, 74

G
Galega (goat's rue) 21, 24, 25, *25*; *G.
× hartlandii* 'Alba' **Z4** *27*; *G. orientalis*
Z5 *72, 73*
Galtonia candicans **Z7** *27*
Gazania 58; *G.* 'Daybreak Bronze' **Z10**
38
Genista aetnensis (Mount Etna broom)
Z8 *98*
Geranium (cranesbill) *12, 77, 97, 102,
107, 109*, 111, 115, *121, 123,* 132; *G.*
'Ann Folkard' **Z5** *4, 102, 121*; *G.
clarkei* 'Kashmir Purple' **Z4** *110, 114,
115*; *G.* 'Johnson's Blue' **Z4** *83*; *G.
kishtvariense* **Z5** *121*; *G. macrorhizum*
'Bevan's Variety' **Z4** *108, 110*; *G.
× magnificum* **Z4** *17, 27, 67, 73, 103,
107, 108,* 114; *G. orientalitibeticum*
Z5 *108*; *G. × oxonianum* 'Wargrave
Pink' **Z5** *108*; *G. pratense* **Z4** *107*;
G.p. 'Mrs Kendall Clark' *108*, 114;
G.p. 'Striatum' 120, *121*; *G.
procurrens* **Z6** 120, *121*; *G. psilostemon*
Z4 *67, 68, 72, 73, 107, 108,* 111,
114; *G.p.* 'Bressingham Flair' *108*; *G.*
'Russell Prichard' **Z6** 102; *G.* 'Salome'
Z5 *121*; *G. sanguineum* **Z4** *32*; *G.s.*
var. *striatum* **Z4** *108*; *G. sylvaticum*

'Mayflower' **Z4** *81*; *G. wallichianum* 'Buxton's Variety' **Z4** 120
geranium (SEE ALSO *Pelargonium*) 58, 79
Gesneriaceae (gesneriads) *119, 123*
Geum 99; *G.* 'Borisii' **Z5** *49*; *G. chiloense* 'Fire Opal' **Z5** *130*; *G.c.* 'Mrs G. Bradshaw' **Z5** *130*
Giverny, France 56
Gladiolus byzantinus **Z7** 68
Gleditsia triacanthos 'Sunburst' **Z3** *99*
goat's rue (*Galega*) **Z4** *21*
goldenrod 34, 85
grape, miller's **Z5** *117*
grape hyacinth (*Muscari*) 102, 137
grasses 12
Great Dixter, East Sussex, Long Border *4, 11*, **96-105**
Gypsophila 86

H

Haberlea rhodopensis **Z6** *119*, 123
Ham House, Surrey 127
Hampton Court, Surrey: Pond Garden 30
Harcourt, Earl of 6
Hardwick Hall, Derbyshire, herb garden 128; West Court borders 12, 95, **126-137**
hazel, purple SEE *Corylus maxima* 'Purpurea'
Healing, Mr and the Hon. Mrs Peter *4*, **86-95**
Hedera helix **Z5** *119*; *H.h.* 'Goldheart' *119*
hedges **16-20**; beech h. 109; yew h. 16, *17, 21, 49, 54, 56, 59, 59*, 88, 102
Hedgleigh Spring, Pennsylvania 10
Helenium 24, 34, 103; *H.* 'Butterpat' **Z4** *81*; *H.* 'Moerheim Beauty' **Z4** *81*, 88, *99, 105*; *H.* 'Riverton Beauty' **Z4** *27, 34, 27, 81*; *H.* Sunshine hybrids **Z4** 88; *H.* 'Wyndley' **Z4** *80, 81*, 88
Helianthus 43; *H.* 'Capenoch Star' **Z5** *88*; *H.* 'Loddon Gold' **Z5** *130*; *H. salicifolius* (syn. *H. orgyalis*) **Z6** 42
Helichrysum bracteatum (syn. *Bracteantha bracteata*) *38, 39*; *H.b.* 'King Size Silvery White' *38, 43*
Heliotropium (heliotrope) 38, 56; *H.* 'Chatsworth' *88*; *H.* 'Marine' *57*, 88
Hellyer, Arthur (1902-92) 9
Heloniopsis orientalis var. *breviscapa* **Z7** *119*
Hemerocallis (daylily) *32*, 46, 48, 73, 77; *H.* 'Dorothy McDade' **Z4** *81*; *H. fulva* 'Flore Pleno' **Z3** *45, 49*; *H.f.* 'Green Kwanso' **Z3** 45; *H. lilioasphodelus* (*H. flava*) **Z3** *77, 81, 82, 83, 99*, 102, 103; *H.* 'Spanish Gold' **Z4** *81*; *H.*

'Stafford' **Z4** *81*
hemlock, Canadian **Z3** 20
Hepatica 6
herbaceous plants 120, 128
herbicides 83; translocated h. 83, *129*
Heslington (The Manor House), York *12*, **106-115**
Hesperis matronalis (sweet rocket) **Z4** 69
Het Loo, Appeldoorn, Netherlands *7, 7*
Heuchera americana **Z4** 48; *H. micrantha* 'Bressingham Bronze' **Z4** 132; *H.m.* 'Palace Purple' **Z4** 48, *49, 130, 131*, 132
Hibiscus syriacus **Z6** 10; *H.s.* 'Oiseau Bleu' (*H.s.* 'Blue Bird') *56*
Hidcote Manor, Gloucestershire 6; Red Borders 11, **44-53**, *93*
Hobhouse, Penelope 82
Hocking, Stephen 86
Hoheria 98
holly (SEE ALSO *Ilex*), American **Z6** 20; European h. **Z7** 20; h. hedges 20; Highclere h. **Z6** 20
hollyhock **Z3** 25
Holmes, Philip 38
hornbeam hedges 20, 53
Hosta *73, 108, 109*, 114, 131; *H.* 'Blue Lotus Leaf' **Z3** 115 *H. fortunei* var. *hyacinthina* **Z3** 115, *115, 137*; *H.* 'Krossa Regal' **Z3** 131; *H. montana* 'Aureomarginata' **Z3** *137*; *H. sieboldiana* var. *elegans* (syn. *H.s.* var. *glauca*) **Z3** *99, 108*, 114; *H.* Tardiana Group 115 *H. ventricosa* **Z3** *67*
Houttuynia cordata 'Flore Pleno' **Z3** *119*; *H.c.* 'Chameleon' **Z3** *99, 104*
Howard, Lady Blanche 128
hyacinth 6, 137
Hydrangea 102, 136; *H. arborescens* 'Annabelle' **Z3** *98, 105*; *H. involucrata* 'Hortensis' **Z7** *124*; *H. macrophylla* 'Mme Emile Mouillère' **Z6** *119*; *H.m.* 'Mariesii' **Z6** *99*; *H. paniculata* **Z4** *119*; *H. aspera* subsp. *sargentiana* **Z8** *118*
Hypericum kouytchense **Z6** *130*

I

Ilex (holly) × *altaclerensis* **Z7** 20; *I. × a.* 'Golden King' **Z7** *99*, 102; *I. aquifolium* **Z7** 20; *I.a.* 'Scotica' **Z7** 20; *I.a.* 'Silver Milkmaid' **Z7** *108, 115*; *I. crenata* 'Convexa' **Z5** 20; *I. opaca* **Z6** 20
Impatiens balsamifera (balsam) 61
Inula magnifica **Z4** *97*, 98, *130*
Iris 10, 64, *67*, 68, 73, 102, *108, 109*, 110, 115, *115*; bulbous i. **Z6** 6; *I. sibirica* **Z4** *108*; *I.s.* 'Keno Gami' *67*;

I. 'Sissinghurst' **Z4** *67*, 69
island beds 7, *9*, 10
ivy (*Hedera*) *118*, 120

J

Jacob's ladder **Z4** *108*
Jekyll, Gertrude (1843-1932) 6, 8, 9, 11, 21, 27, 34, 37, *54*, 56, *57, 63*, 77, 78, 79, *80*, 86, 89, 97, 98, *98*, 102, 111, 128, 131
John, Augustus (1878-1961) 45
Johnson, George W. (1802-86) 16
Johnston, Colonel Lawrence (1871-1958) 11, 45, 46, 48, 52, 53
Juniperus (juniper), Irish **Z3** 94; *J. scopulorum* 'Skyrocket' **Z4** *108*

K

Kaye, Reginald (1902-92) 121
Kemerton (The Priory), Hereford and Worcester, Top Border *4*, **86-95**
Knautia macedonica (syn. *Scabiosa rumelica*) **Z5** *72*, 88, *110*
Kniphofia (red hot poker) 10, 25, 94, 98, 102, 105, 131; *K.* 'Green Jade' **Z5** *88*; *K.* 'Lord Roberts' **Z5** *99*; *K.* 'Samuel's Sensation' **Z5** *130*; *K. uvaria* 'Nobilis' **Z5** *88*, 94
Kreutzberger, Sibylle 64, 68, 74

L

labels *69, 72*
lamb's ears SEE *Stachys byzantina*
landscape style 7
larkspur 25
Lathyrus 120; *L. chloranthus* 120; *L. grandiflorus* **Z6** *103*
Lavandula (lavender) *108*; *L. angustifolia* 'Hidcote' **Z6** *63, 67*, 74; *L. stoechas* **Z8** *108*
Lavatera 40, 58; *L.* 'Barnsley' **Z8** *79*, 82, 88, 89, 94, 95; *L.* 'Loveliness' *61*; *L.* 'Rosea' **Z8** 82
lawns 52, 53
Leymus arenarius (syn. *Elymus a.*) **Z4** 104
Liatris spicata **Z4** *63, 67*, 68
Ligularia przewalskii **Z4** *27*
Ligustrum (privet) *119, 124*; *L. ovalifolium* 'Argenteum' (silver p.) **Z6** *119*, 120; *L.* 'Vicaryi' **Z6** 104
lilac 47, 98, 103
Lilium (lily) 6, *80*, 83, 132; Asiatic hybrid l. 132; *L.* 'Destiny' **Z3** *81*; Dr North's hybrids 132; *L. henryi* **Z4** 132; *L. lancifolium* (syn. *L. tigrinum*) 'Flore Pleno' (double tiger l.) **Z3** *130*, 132; *L.l.* var. *fortunei* **Z3** *81*; *L. longiflorum* **Z8** 83; Madonna l. **Z4** 56, *57*; *L.* 'Orestes' **Z4** 132; *L. pardalinum*

(leopard l.) **Z5** *49*
Linaria 124
Lindsay, Mrs Norah (1876-1948) 46
Lloyd, Christopher *4*, 68, **97-105**
Lloyd-Jones, Kitty 54, *54*, 56, 59-61
Lobelia 46, *57*, 123; *L.* × *gerardii* 'Vedrariensis' **Z4** 68, *118*; *L.* × *speciosa* **Z4** 45, 52; *L.* × *s.* 'Bees' Flame' *49*, 52; *L.* × *s.* 'Cherry Ripe' *49*, 52; *L.* × *s.* 'Dark Crusader' 45; *L.* × *s.* 'Queen Victoria' 52; *L.* × *s.* 'Will Scarlet' 52
Lonicera 124; *L.* × *americana* **Z6** 98, 103; *L.* × *italica* **Z6** 103; *L. nitida* 'Baggesen's Gold' **Z7** *98*, 120
Loudon, John Claudius (1783-1843) 16, 20, 29
Lunaria rediviva **Z6** *119*
lupin **Z3** 89, *107, 108, 109*, 110, 115; l. 'Blue Jacket' **Z3** *67*
Lutyens, Sir Edwin (1869-1944) 97, *97*
Lychnis chalcedonica **Z4** *32*, 35, *81*, 98, *99*; *L. coronaria* **Z4** *25, 27*
lyme grass (*Elymus, Leymus*) *54*, 56, *56*
Lysimachia ciliata **Z3** *25*; *L. ephemera* **Z7** *21*
Lythrum 17, 79; *L. salicaria* 'Feuerkerze' **Z4** *32*; *L. salicaria* 'Robert' **Z4** *63, 67*

M

Macleaya (plume poppy) *21, 25, 27*, 82; *M. cordata* **Z4** *27, 29, 32*, 34
Magnolia grandiflora **Z7** *80*
Malva (mallow) 74, 95; *M. alcea* **Z4** 88; *M. moschata* **Z4** *27*; *M. sylvestris* var. *mauritiana* *67*, 69, *73*, 74, 89, 95
manganese toxicity 74
manure 34, 124
maple, Japanese **Z6** 47
marigold, African 38, 56, 58, *61*
marigold, French 38, 56, 58, *61*
marigold, pot 25
Marrubium incanum **Z3** *108*
Mason, William (1725-97) 7
Massen, Rob 38
Masters, David 37, 41, 42
Mawson, Thomas H. (1861-1933) 46
Messel, Ludwig 37
Michaelmas daisy (SEE ALSO *Aster novi-belgii*) **Z4** 64, 68
mildew 35, 74, 125
Mimulus *129*; *M. puniceus* **Z9** *132*
'mingled' style 11, **29-35**
Miscanthus 48, 131; *M. sinensis* 'Gracillimus' **Z4** *45*, 46, 49; *M.s.* 'Silberfeder' (*M.s.* 'Silver Feather') **Z5** *130*; *M.s.* 'Zebrinus' **Z6** *98*; variegated m. 105
Moluccella laevis 27
Monarda 74, 83, 88, 89, 95, 103; *M.*

'Beauty of Cobham' **Z4** *63, 67; M.* 'Cambridge Scarlet' **Z4** *32,* 35, 83, *88; M.* 'Mrs Perry' **Z4** *88, 89; M.* 'Prärienacht' (*M.* 'Prairie Night') **Z4** *67; M.* 'Vintage Wine' **Z4** *67*
Monet, Claude (1840-1926) 56
Montacute, Somerset *7, 21,* 128, 129
montbretia *17, 24*
Mottisfont Abbey, Hampshire, double borders *10,* 88
mulberry 85
mulching 124
Munstead Wood, Surrey *78*
Muscari (grape hyacinth) 102, *137*

N
Narcissus 137
nasturtium (*Tropaeolum*) *98,* 104
Nepeta (catmint) *54, 110; N. sibirica* **Z4** *72; N.* 'Six Hills Giant' **Z4** *32, 34, 35,* 137
Newstead Abbey, Nottinghamshire 15
Nice, Cecil 37
Nicolson, Sir Harold (1886-1968) 63-65, 68, *74*
Nicholls, Paul 45, 46, 48, 52
Nicotiana (tobacco flower) 56, 61, 123; *N. alata; N.* 'Domino Pink' *38; N. sylvestris 58*
Nigella 121
Nuneham Park, Oxfordshire *7*
Nymans, Sussex, Summer Borders 9, 12, **36-43**

O
oak, holm or ilex **Z7** 47, *53*
Oehme, Wolfgang 10
Omphalodes cappadocica **Z6** 102
Onopordum nervosum **Z8** 21, 24, *27,* 73, *79,* 85, *108,* 114
Ophiopogon 119
orach, purple 15, *17, 24, 27, 89, 93*
orientation of borders *38, 49*
Origanum 120
Osteospermum 118

P
Packwood House, Warwickshire, Sunken Garden *30;* Yellow Border 9, 11, **28-35;** Yew Garden *30*
Paeonia (peony) 6, *107, 108, 109,* 110, 115, *115*
Page, Russell (1906-85) 56
palissades à l'italienne 53
pampas grass **Z7** *131*
Panicum virgatum 'Rubrum' **Z5** *130*
pansy 25
Papaver (SEE ALSO poppy); *P. dubium 121; P. orientale* (oriental poppy) **Z4**

'Beauty of Livermere' (syn. *P.o.* 'Goliath') *49; P.o.* 'Cedric's Pink' *108; P.o.* 'Graue Witwe' (*P.o.* 'Grey Widow') *27; P.o.* 'Sultana' *107, 108,* 111; *P. somniferum 27; P. spicatum* (syn. *P. heldreichii*) **Z4** *130*
Parsons, Beatrice (1870-1955) 37
Parthenocissus inserta **Z3** 103
Patrinia gibbosa **Z5** *119*
pea SEE *Lathyrus*
Pelargonium 52, *57; P.* 'Arley' *88; P.* 'Century Scarlet Eye' *38; P.* 'Général Championnet' *49,* 52
Penstemon 93, *108,* 110, *127,* 136; *P.* 'Alice Hindley' **Z8** *89,* 110, 136; *P.* 'Andenken an Friedrich Hahn' (*P.* 'Garnet') **Z6** *7,* 83, *88,* 93, *131,* 136; *P.* 'Burgundy' **Z8** 136; *P.* 'Mother of Pearl' **Z8** 93; *P.* 'Papal Purple' **Z8** 136; *P.* 'Port Wine' **Z8** 136; *P.* 'Rich Ruby' **Z8** 93, 136; *P.* 'Schoenholzeri' (*P.* 'Firebird') **Z6** 93, *131, 132,* 136; *P.* 'Stapleford Gem' (*P.* 'Sour Grapes' misapplied) **Z8** 110; *P.* 'White Bedder' (syn. *P.* 'Snowstorm') **Z8** 93
peony SEE *Paeonia*
perennials, herbaceous 8, *12;* tender p. 8, 10, 11, 12, 38, 52, 58, **126-137**
Perilla frutescens var. *nankinensis 38, 39,* 41
Perovskia 54, 56, *56,* 57
Persicaria (syn. *Polygonum*) *amplexicaulis* 'Firetail' **Z5** *79, 81; P. bistorta* 'Superba' **Z4** 68
Petunia 38; *P.* 'Mirage White' *38; P.* 'Purple Defiance' *67; P.* 'Sugar Daddy' *38, 43*
Phacelia 25
Phlomis fruticosa **Z8** *98, 99,* 102
Phlox 24, *25,* 34, *45,* 56, *56, 61,* 68, *79, 80,* 83, *97, 99,* 102, 103; *P. carolina* 'Bill Baker' **Z4** *119; P. drummondii* 38, *58; P. maculata* **Z4** *119; P.m.* Alba; *P.m.* 'Alpha' *97; P. paniculata* **Z4** *17, 27, 32,* 42, *57, 81,* 98; *P.p.* 'Amethyst' *131; P.p.* 'Brigadier' *130; P.p.* 'Cool of the Evening' *67; P.p.* 'Frau Antoine Buchner' *82; P.p.* 'Mia Ruys' *82; P.p.* 'Starfire' *88, 95*
Photinia 118, 125
Phuopsis stylosa **Z5** *108,* 110
Phygelius capensis **Z8** 88
Phyllitis 119; P. scolopendrium (syn. *Asplenium* s.) Crispa Group **Z4** *119*
Picton, Percy 120
pinks *12,* 25
Pinus mugo **Z3** 46, *47*
plates bandes 7, 7

Platycodon grandiflorus **Z4** *67,* 68
plum, purple SEE *Prunus cerasifera* 'Pissardii'
polyanthus, Cowichan **Z4** 48
Polygonatum × *hybridum* 'Striatum' **Z4** *119*
Polygonum SEE *Persicaria*
polymer, water-retentive 39
Pontrancart, Le, Dieppe, France, 12; Old Garden **54-61;** blue and white garden *54,* 56, *59;* cosmos border *59;* dahlia borders *59;* main borders *59, 61;* pink and red border 60, *61*
poplar, white **Z4** *61*
poppy (SEE ALSO *Papaver*) 15, *16, 21, 25, 27,* 110, 114; oriental p. **Z4** *98, 99;* o.p. 'Cedric's Pink'; Shirley p. *25,* 39
poppy, Californian 38
poppy, plume SEE *Macleaya*
Potentilla 'Gibson's Scarlet' **Z5** *32,* 35, *49, 130, 131; P. recta* var. *pallida* **Z4** *29, 32, 34, 35*
Powis Castle, Powys, Wales 68, 85, 136
primrose 6, 102
privet SEE *Ligustrum*
propagation 93
pruning 124
Prunus cerasifera 'Pissardii' (purple plum) **Z4** 48, *118, 120, 123; P. spinosa* 'Purpurea' (purple sloe) **Z5** *47,* 48, *49, 131, 132; P.* × *subhirtella* 'Autumnalis' **Z6** *108, 120*
Puckering, Norman 108
Pulmonaria 48, 102, *119; P. rubra* **Z5** *49*

Q
queen of the prairie SEE *Filipendula rubra*

R
Ramonda myconi **Z6** *118, 119,* 123
Ranunculus 6
Rea, John (d. 1681) 6, 7, *7,* 16
red hot poker SEE *Kniphofia*
red spider 40
Reiss, Phyllis (d. 1961) *7,* 82, 83, 129
Repton, Humphry (1752-1818) 7, 16
Rheum palmatum 'Atrosanguineum' **Z5** *49, 130*
rhizoctonia 40
Rice, Graham 54
Ricinus 94; *R. communis* 'Gibsonii' *88,* 93
Robinia pseudoacacia 'Frisia' **Z3** *118,* 120
Robinson, William (1838-1935) 8, 37, 86, 120
rocket, double **Z4** 6
Rodgersia 119; R. pinnata **Z5** *119*
Rosa (rose) *12,* 31, *32,* 35, *49,* 102, 108,

108, 109, *109,* 114, 115, 127, 128, 136; *R.* 'Ballerina' **Z6** *98; R.* 'Carmenetta' **Z4** *108, 115; R.* 'Cerise Bouquet' *107, 110; R.* 'Charles de Mills' **Z6** *108; R.* 'Chinatown' **Z6** 136, *137; R.* 'Complicata' **Z6** *108; R.* 'Fantin-Latour' **Z6** *12; R.* 'Florence Mary Morse' **Z6** *97,* 102; *R.* 'Fred Loads' **Z6** *130,* 136; *R.* 'Frensham' **Z6** *45, 49; R.* 'Frühlingsgold' **Z5** *108; R.* 'Geranium' **Z6** *45, 49, 63,* 65, *67,* 72, *74,* 88, *93,* 94, *131; R. glauca* (syn. *R. rubrifolia*) **Z2** *88; R.* 'Golden Wings' **Z6** *108; R.* 'Golden Years' **Z6** *88; R.* 'Iceberg' **Z6** 56, *61; R.* 'Marlena' **Z6** *136; R.* 'Matangi' **Z6** *108, 109; R. moyesii* **Z6** *64,* 68, 102; *R. multibracteata* **Z7** *110; R.* 'News' **Z6** *72; R. nutkana* 'Plena' (*R. californica* 'Plena' misapplied) **Z6** *108, 108, 109; R.* 'Orange Triumph' **Z6** *49,* 52; *R.* 'Penelope' **Z6** *108; R.* 'Rosemary Rose' **Z6** *45; R. setipoda* **Z6** 102; *R.* 'The Times' **Z6** *136; R. xanthina* 'Canary Bird' **Z6** *108, 108, 115; R.* 'Yesterday' **Z6** *108; R.* 'Zéphirine Drouhin' **Z6** *27,* 98
rose replant disease 129
Rubus arcticus **Z1** *119; R. phoenicolasius* **Z6** *98*
Rudbeckia 17, 56, 88; *R. fulgida* var. *deamii* **Z4** *88; R.* 'Herbstsonne' (*R.* 'Autumn Sun') **Z3** *38, 81,* 88; *R. maxima* **Z6** *80, 81; R. mollis* (gloriosa daisy) *130*
rust disease 74
Ruta graveolens (rue) 'Jackman's Blue' **Z5** *77*
ryegrass, dwarf 53

S
Sackville-West, Vita (1892-1962) *4,* 11, 63-65, 68, *74,* 128
Sales, John 63, 85
Salix alba var. *sericea* (silver willow) **Z2** 102, 103
Salvia 7, 11, 52, *80; S. bulleyana* **Z5** *32, 34, 35, 38; S. coccinea* **Z9** 39; *S.c.* 'Lady in Red' *61; S. elegans* **Z9** *45; S. farinacea* **Z9** 56; *S. involucrata* 'Bethellii' **Z9** *132; S.i.* 'Boutin' **Z9** *131, 132; S. lycioides* **Z9** *67, 74; S. microphylla* var. *neurepia* **Z9** *49; S. nemorosa* 'Ostfriesland' (*S.n.* 'East Friesland') **Z5** *27, 77, 81; S. officinalis* Purpurascens Group **Z6** *49; S. patens* **Z9** 25, 54, 56, *57; S. sclarea* var. *turkestanica* **Z5** *27,* 31, *32, 89; S.* × *superba* **Z5** *67,* 68, 73, *80, 81,* 85,

88, 95, 99, 102, 103; *S. uliginosa* **Z9** 99, 104; *S. viridis* (syn. *S. horminum*) 'Bluebeard' 67

Sambucus racemosa 'Plumosa Aurea' (golden elder) **Z4** 103

Sandon, Staffordshire 16

saxifrage, meadow, double **Z5** 6

Saye and Sele, Lord and Lady 12

Scabiosa (scabious) 'Butterfly Blue' **Z5** 27

scabious, giant **Z3** 17

Schizostylis coccinea **Z6** 99

Schwerdt, Pamela 64, 68, 74

Sedum 77; *S.* 'Herbstfreude' (*S.* 'Autumn Joy') **Z3** 81; *S.* 'Ruby Glow' **Z4** 131; *S. spectabile* 'Brilliant' **Z4** 27, 131; *S. telephium* 'Arthur Branch' **Z4** 120; *S.t.* subsp. *maximum* 'Atropurpureum' **Z4** 88, 111, 131; *S. t.* subsp. *ruprechtii* **Z4** 118; *S.* 'Vera Jameson' **Z4** 67

seed sowing 39

Senecio cineraria (dusty miller) **Z8** 54, 57, 98; *S. doria* **Z6** 98, 105; *S. viravira* **Z9** 98

Sidalcea 34, 108; *S. candida* **Z5** 24; *S.* 'Loveliness' **Z5** 32

Silphium perfoliatum **Z3** 42

Sissinghurst Castle, Kent 6, 11, 82, 136; Purple Border *4*, **62-75**, 102

Sisyrinchium striatum **Z7** 10

Skimmia japonica **Z7** 119

sloe, purple SEE *Prunus spinosa* 'Purpurea'

slugs 74, 120, 125

Smith, George **107-115**

smoke bush SEE *Cotinus coggygria*

Smythson, Robert (*c*1535-1614) 127

snapdragon SEE *Antirrhinum*

snowdrop 102

soap, horticultural 40

Solanum crispum **Z8** 27

Solidago 24, 27, 32, 88, 95; *S.* 'Golden Gates' **Z4** 81; *S.* 'Lemore' **Z4** 32; *S.* 'Strahlenkrone' (*S.* 'Crown of Rays') **Z4** 88

sparse planting 6

Spartium junceum (Spanish broom) **Z8** *129*, 130, 132

Stachys byzantina **Z4** 10, 54, 57, 108, 114; *S. coccinea* **Z9** 130, 132, 133; *S. macrantha* **Z4** 67, 68

staking *30*, **31**, 34, 41, *41*, 52, 67, **72-73**, *72*, 79, 80, 82, 83, 94

sterilization, soil 83, *129*, 129

stock, double 6

Stonecrop, New York 12

strawflower, Australian SEE *Helichrysum bracteatum*

Sunningdale Nurseries, Surrey 108

sweet pea 'Noel Sutton' 65, *67*

sweet rocket (*Hesperis matronalis*) **Z4** 69, 69

Synthyris stellata **Z7** 119

T

Tamarix (tamarisk) 99, 102; *T. ramosissima* **Z3** 99

Tanacetum ptarmiciflorum **Z9** 54, *57*

tarsonemid mite 74

Taxus baccata **Z6** 20; *T cuspidata* **Z5** 20; *T. × media* **Z5** 20

teasel 98, 99

Thalictrum aquilegiifolium **Z5** *32*; *T. delavayi* (syn. *T. dipterocarpum*) **Z5** 63, 64, 67, 68, 73; *T. flavum* subsp. *glaucum* **Z6** 7, 27, 32, 81

Thomas, Graham Stuart 10, *10*, 45, 47, 48, 52, 53, 54, 56, 63, 64, *65*, **77-85**, 108, 109, 111, 137

Thorold, Sir George 97, 98

Tintinhull, Somerset *7*, 83, 129; Pool Garden 82

tobacco flower SEE *Nicotiana*

Tolmiea menziesii 'Taff's Gold' **Z7** 119

Trachymene coerulea (syn. *Didiscus coeruleus*, blue lace flower) 56, *57*, 58

Treasure, John (1911-93) 111

Tricyrtis 121; *T. formosana* **Z5** 119

triforine fungicide 74

Trillium 121

Tsuga canadensis **Z3** 20

Tulipa (tulip) 6, 7, 48, 52, 69, 72, 83, 89, 102, *103*, 110, 120; *T.* 'Absalom' **Z4** *103*; *T.* 'Angélique' **Z4** 110; *T.* 'Attila' **Z4** 72; *T.* 'Black Parrot' **Z4** 110; *T.* 'Blue Parrot' **Z4** 69, 72, 110; *T.* 'Dairy Maid' **Z4** 69, 72; *T.* 'Dyanito' **Z4** 47, 52; *T.* 'Greuze' **Z4** 47, 72; *T.* 'Groenland' **Z4** 110; *T.* 'Olaf' **Z4** 52; parrot t. *7*; *T.* 'Queen of Night' **Z4** 47, 48, 110, *110*; *T.* 'Red Shine' **Z4** 47, 49, 52; *T.* 'Ruby Red' **Z4** 48

turf 52, 53

Twickenham House, Surrey 127

U

Ulmus 'Dicksonii' **Z5** 98, 102, *118*, 120

Underwood, Mrs Desmond 64, 111

Upton House, Warwickshire 54

V

valerian 34

Vancouveria hexandra **Z5** 119

Van Sweden, James 10

variegation 115, 120, 132

Verbascum 25, 83, 98, 99, 132; *V. nigrum* **Z5** *105*; *V. olympicum* **Z6** 102; *V.* 'Vernale' **Z5** 81, 85

Verbena 52, *57*, 111, 129, 132; *V.* *bonariensis* **Z9** 97, 98, 99, 105, *105*; *V.* 'Kemerton' **Z9** 130, 131, 132; *V.* 'Lawrence Johnston' **Z9** 49, 52, 88; *V. rigida* (syn. *V. venosa*) **Z9** 79; *V.* 'Silver Anne' **Z9** 67

vermiculite 39

Veronica spicata **Z4** 27

Veronicastrum (syn. *Veronica*) *virginicum album* **Z4** 38

Viburnum 124; *V. opulus* 'Roseum' (syn. *V.o.* 'Sterile') **Z4** 108, *108*

vine weevil 40

Viola (viola, pansy, violet, violetta) 102, *108*; violet 102; violetta 64; *V.* 'Huntercombe Purple' **Z6** 131

Vitis vinifera 'Incana' (miller's grape) **Z5** 117, 120; *V.v.* 'Purpurea' **Z5** 67

W

wallflower **Z7** 6, *30*, 34, 35, 48, 69, *69*, 72, 93

Warburton, Sir Peter 15

Wardian case 8

watering 40

weeds 42, 93

whitefly 40

Willmott, Ellen Ann (1858-1934) 86, 120

willow SEE *Salix*

willowherb, rosebay 120

wilt, clematis 69

Wimbledon House, Surrey 127

Winthrop, Mrs Gertrude (1845-1926) 46

Wisley, Royal Horticultural Society's Garden, Surrey 24

woodland plants *119*, 120

Y

yew, European **Z6** 20, 88, *131*; fastigiate golden y. 94; y. finial *38*, *41*; y. hedges 16-20, *16*, *17*, *21*, *27*, *49*, *54*, *56*, *59*, *59*, 86, *88*, 102; topiary y. *7*

Yucca 78; *Y. flaccida* 'Ivory' **Z5** 130, 131

Z

Zinnia 58

HARDINESS ZONES

The hardiness zone ratings given for each plant – indicated in the index by the letter 'Z' and the relevant number – suggest the appropriate minimum temperature a plant will tolerate in winter. However, this can only be a rough guide. The hardiness depends on a great many factors, including the depth of a plant's roots, its water content at the onset of frost, the duration of cold weather, the force of the wind, and the length of, and the temperatures encountered during, the preceding summer. The zone ratings are based on those devised by the United States Department of Agriculture. Zone ratings are allocated to plants according to their tolerance of winter cold in the British Isles and Western Europe. In climates with hotter and/or drier summers, as in Australia and New Zealand, some plants will survive colder temperatures and their hardiness in these countries may occasionally be one, or even rarely two, zones lower than that quoted.

CELSIUS	ZONES	°FAHRENHEIT
below -45	1	below -50
-45 to -40	2	-50 to -40
-40 to -34	3	-40 to -30
-34 to -29	4	-30 to -20
-29 to -23	5	-20 to -10
-23 to -18	6	-10 to 0
-18 to -12	7	0 to 10
-12 to -7	8	10 to 20
-7 to -1	9	20 to 30
-1 to 4	10	30 to 40
above 4	11	above 40

Acknowledgments

AUTHOR'S ACKNOWLEDGMENTS

I am grateful to all the staff at Frances Lincoln Limited who have worked so hard, so quickly and with their customary high standards of production to create *Best Borders*. Especial thanks go to Penelope Miller, Louise Tucker, Caroline Hillier, Erica Hunningher and Anne Fraser, and to Frances Lincoln for courageously giving a novice writer and photographer such a wonderful opportunity for a first book.

The work of photographing and researching each of the featured gardens has been made a pleasure by the unfailing helpfulness and hospitality of every garden owner, manager and gardener consulted. My particular thanks go to the following: at Arley, Lady Ashbrook, Mr and the Hon. Mrs Charles Foster, Tommy Acton and Paul Cook; at Packwood, Fred Corrin; at Nymans, David Masters and Rob Massen; at Hidcote, Paul Nicholls; the owner and Monsieur A. Allix of Le Pontrancart; at Sissinghurst, Pamela Schwerdt, Sibylle Kreutzberger and Sarah Cook; at Cliveden, Graham Stuart Thomas and Philip Cotton; at the Priory, Kemerton, the Hon. Mrs Peter Healing, Charles Brazier and Stephen Hocking; Christopher Lloyd at Great Dixter; George Smith at the Manor House, Heslington; Mrs Anne Dexter in Oxford and, at Hardwick, Robin Allan. I am grateful to Joe Eck for his suggestions of North American gardens with fine borders.

It would scarcely have been possible to produce such a book without including a substantial proportion of National Trust gardens. Their historical importance, variety and high standards of maintenance, their continuity with the past while still being revitalized by new planting within historical guidelines, make them an inspiration to all. I am grateful to the Trust's Regional Directors who have allowed me to pester hard-pressed Head Gardeners with innumerable questions, to Chief Gardens Adviser John Sales for his encouragement and to Gardens Consultant Graham Thomas for a lengthy interview setting out his influential views on border planting. My thanks go also to all the Trust gardeners who have played an essential role in maintaining the gardens' quality and individuality. I am grateful also to my former colleague Isabelle Van Groeningen for discussing with me her current research on borders.

The photographs were almost all taken using a Canon T90 camera and Fuji Velvia film, supplied by Jessops Cheltenham Branch and developed by Central Photographic Services, Cheltenham. I am grateful to the staff of both for their courteous, speedy and excellent service.

PUBLISHERS' ACKNOWLEDGMENTS

The Publishers would like to thank the following people for their help in producing this book: Biggles, Jo Christian, Barbara Fuller, Studio Gossett, Diana Loxley, Mr Lyddon, Anthea Matthison, Maggi McCormick, Auriol Miller, Annabel Morgan, Barbara Wood

Plan artwork   Pond and Giles
Editor   Penelope Miller
Art Editor   Louise Tucker
Production   Annemarieke Kroon

Editorial Director   Erica Hunningher
Art Director   Caroline Hillier
Production Director   Nicky Bowden